ARROYO CENTER

C000007701

Quality of Life at Contingency Bases

Managing Sustainment Community Support Requirements

John E. Peters, Elvira N. Loredo, Mary E. Chenoweth, Jeff Broughton, Andrea A. Golay

Prepared for the United States Army

For more information on this publication, visit www.rand.org/t/RR1298

Library of Congress Cataloging-in-Publication Data is available for this publication.

ISBN: 978-1-9774-0186-1

Published by the RAND Corporation, Santa Monica, Calif.

© Copyright 2018 RAND Corporation

RAND® is a registered trademark.

Support RAND

Make a tax-deductible charitable contribution at
www.rand.org/giving/contribute

www.rand.org

Preface

This report documents research and analysis conducted as part of a project entitled *Developing Policies for Supportable Quality-of-Life Programs at Contingency Bases*, sponsored by the Deputy Chief of Staff of the U.S. Army for Logistics. The purpose of the project was to characterize the demand and to identify options for managing quality-of-life goods and services delivered for the benefit of deployed forces operating from base camps and that can produce significant demand for resources, including fuel, support personnel, trucks, and other distribution assets.

The Project Unique Identification Code (PUIC) for the project that produced this document is HQD146658.

This research was conducted within RAND Arroyo Center's Forces and Logistics Program. RAND Arroyo Center, part of the RAND Corporation, is a federally funded research and development center (FFRDC) sponsored by the United States Army.

RAND operates under a "Federal-Wide Assurance" (FWA00003425) and complies with the *Code of Federal Regulations for the Protection of Human Subjects Under United States Law* (45 CFR 46), also known as "the Common Rule," as well as with the implementation guidance set forth in DoD Instruction 3216.02. As applicable, this compliance includes reviews and approvals by RAND's Institutional Review Board (the Human Subjects Protection Committee) and by the U.S. Army. The views of sources utilized in this study are solely their own and do not represent the official policy or position of DoD or the U.S. Government.

Contents

Figures

Tables

Summary

Leaders and soldiers alike want to have the best quality of life (QoL) possible, given the circumstances in the theaters of operation to which they deploy. Logisticians, in turn, are keenly aware of how choices related to QoL levels influence fuel, water, food, and overall transportation requirements in a theater. Recent experience in Afghanistan and Iraq demonstrates that long-term deployments can lead to an escalating cycle of "improvements to the foxhole" that ultimately drive costs, add trucks to convoys, and put additional lives at risk.

Definitions of *QoL* vary, depending on an organization's mission. Engineers who design contingency bases tend to define it in terms of engineering standards (e.g., *n* toilet seats and shower heads per *y* population). Service organizations, such as the Army and Air Force Exchange Service (AAFES), tend to think of QoL in terms of the relative abundance of goods and services they can provide. The morale, welfare, and recreation organization thinks of its contribution to soldier QoL in terms of the number of internet connections it can provide and the quality and variety of recreational and entertainment options it can offer. In this report, we explore the balance between QoL required to support soldiers and the logistics required to establish and sustain QoL on a base camp. To begin, we developed a provisional definition of *QoL* to capture a sense of all of the factors that contribute to soldier well-being:

> QoL at contingency bases is determined by the equitable distribution of material and personal goods and services beyond the essentials necessary for simple survival, to enable a soldier to remain fit to fight despite hardships and deprivations encountered in the course of his or her duties.

In this report, we reference research on the importance of QoL to soldier well-being and soldier perceptions of what is most important. These include billeting, food, communications home (including access to Wi-Fi and Skype), and hygiene (showers, latrines, and laundry). Then, using data from base camps in Afghanistan on fuel requirements and engineering test and evaluation results on electrical and water systems, we begin to establish a relationship between QoL levels and logistics requirements.

From the logistics data, we find that support for billeting is a major driver of non–operations-related fuel requirements and that food and water used for showers, latrines, and laundry are also important determinants of the number of convoys needed to support contingency bases. Therefore, focusing on how decisions are made in relation to billeting, showers, laundry, latrines, and food will have the greatest effect both on the logistic burden that QoL creates and on how soldiers perceive the benefits of QoL.

By doctrine, the base-camp commander is responsible for the base-camp QoL. Army Techniques Publication 3-37.10, *Base Camps*, provides some guidance on how the level of QoL should be determined (Headquarters, Department of the Army [HQDA], 2013a). Army Techniques Publication 3-37.10 states that the levels of base-camp capabilities and the nature of the construction effort required should be commensurate with the anticipated duration of the mission. *Commander and Staff Officer Guide*, Army Tactics, Techniques, and Procedures 5-0.1, provides logical categories for QoL support and services. Annex F of this Army tactics, techniques, and procedures organizes QoL-related commodities and services. Table S.1 provides examples of QoL elements in the material and services and personnel categories.

The logistics burden is generated primarily by the components listed for material and services. However, Army doctrine and policy are silent on how to manage QoL on contingency bases with the goal of controlling or possibly lessening the logistics burden. Lack of a clear link between logistics burden and QoL means that QoL can evolve at contingency bases to a scale that is difficult to support. This evolution reflects long-standing Army practice that soldiers always work to improve their positions. This is reflected in how engineering standards for contingency bases are time-phased, first providing rather austere initial facilities, later temporary facilities, and finally further development of the camp to semi-permanent accommodations.

As stated previously, most QoL demands on the sustainment community come down to fuel, water, and food. In the next section, we summarize options to reduce the logistics burden related to these commodity areas.

Improvements by Commodity Area

Food

Our research finds that—as expected—the fresher the food, the greater its weight. The research team hypothesized a division-level task force of five brigade combat teams and their enablers, or 19,283 personnel, and then estimated their food requirements for a year (21.1 million meals). If fed prepackaged individual meals, ready to eat (MREs), the force would require 12,220 tons of these rations (1,111 M977 heavy expanded mobile tactical truck loads). A B-ration (meals prepared using canned or preserved ingredients) diet would weigh 17,596 tons (1,600 M977 loads). From these data, we calculate a 44-percent convoy premium for B-rations over MREs. An A-ration (meals prepared using fresh, refrigerated, or frozen foods) diet for the force would weigh 19,283 tons (1,753 M977 loads) and a 58-percent convoy premium over MREs. Stated another way, if we assume one soldier casualty for every 20 convoys and 25 vehicles per convoy, this task force can expect one to one and a half additional casualties per year as a result of shipping B-rations or A-rations, respectively, over MREs.

In reality, dining facilities at large forward operating bases feed fresh food, often four meals per day. Packaged rations tend to be consumed during operations and while manning combat outposts where security and lines of communication might be more contested. Con-

Table S.1
Quality-of-Life Elements, by Category

Category	Element
Material or services	Housing, meals and feeding, shower, latrine, and waste removal and recycling
Personnel	Chapel and religious services, internet café, fitness center, AAFES, movie theater, telephone call center, educational center, other vendor services (including local bazaar, post office, and banking), United Service Organizations, and base-camp transit or bus route

sumption and spoilage drive replenishment rates. Members of the 377th Theater Sustainment Command interviewed for an earlier project reported that all of the convoys they operated in Iraq involved delivery of rations. Of course, soldiers must always be fed, no matter where they are stationed, so food deliveries are unavoidable. Further, as our research found, access to fresh food is an important contributor to the soldier's perception of QoL. Further analysis needs to be done to optimize the mix of meals, but clearly the soldier security and security of the lines of communication need to be considered when determining the mix of meals.

Retail restaurants, however, differ from dining facilities or force-provider kitchens because they have an additional water penalty (above that needed for a A-rations) for each meal served. As the parametric estimates in Table S.2 suggest, the potential savings could run to hundreds of tankers' worth of water per camp per year for theaters with water limitations similar to those found in Afghanistan and Iraq.

The sustainment community should work with combatant command (COCOM) logistics staff to develop a policy on retail restaurants and similar concessions on contingency bases. The policy should be grounded in the supportability of water requirements under the conditions prevailing in the theater of operations.

Water

An estimated 75 percent of the water required at contingency bases is utilized for showers. Since 2011, the Army has fielded 54 shower water–reuse systems (SWRSs) to units in Afghanistan. An SWRS combines the tactical water-purification system and hospital containerized batch laundry capabilities to treat and return to use up to 9,000 gallons of water per day. Using just one system at its full capacity can result in saving potentially 3.2 million gallons of water per year.

In addition to the SWRS, the Army is investigating other opportunities, including graywater recovery from laundry and food preparation and similar sources. Table S.3 illustrates the research team's estimates of water savings based on the 2014 U.S. Army troop strength in Afghanistan (32,720 soldiers), assuming a requirement for 60 gallons per person per day and 30-, 40-, and 50-percent reuse.

Energy

Base-camp energy is an area rich in potential savings, both in terms of electricity and in terms of fuel. One potentially very effective fuel-saving option is the selection of soldier housing and work location structures. Many options are available, from general-purpose medium tents to semi-permanent military construction at the other end of the spectrum. In between these

Table S.2
Retail Meals and Additional Water Demand

Camp Population	Camps in Sample	AAFES Restaurant Allocation	Total Restaurants	Water Consumption, in Trucks per Year[a]
1,000–2,000	9	1	9	1,381
2,000–3,000	1	2	2	307
3,000–4,000	2	2	4	614

NOTE: AAFES does not operate every restaurant on base.

[a] Assuming 300 restaurant meals per day.

Table S.3
Estimates of Savings from Water Reuse, in Tanker-Equivalents per Month

Percentage Recovered	Tanker-Equivalents Saved Monthly
30	6,690
40	8,910
50	11,160

two extremes are numerous prefabricated housing solutions. Each of these requires a different amount of fuel to achieve the same climate-control and electrical capabilities. In many cases, the differences are remarkable. One illustration of this effect involves replacing wooden barrack huts (B-huts) with structural insulated panel (SIP) huts (see Appendix D). According to the research team's contractor-provided consumption reports, total monthly contingency base energy consumption for the 16 camps in our data set totals 1.4 million gallons of fuel. If we assume that 90 percent of that goes to living, dining, and office space provided in B-huts, the total would be approximately 1.2 million gallons. Given that SIP huts' energy demand is 60 percent of B-hut demand because of SIP huts' superior insulation, the reduced energy demand would be approximately 504,000 gallons, or an estimated monthly savings of about 42,000 gallons.

Other energy savings found in electricity generation include (1) move from spot generation to grid-based electricity, (2) exchange the Vietnam-era tactical quiet generators for today's modern Army's Advanced Medium Mobile Power Sources (AMMPS) systems. In a direct exchange of tactical quiet generators for AMMPS, such a move would affect 4,826 generators in Afghanistan producing 60 kilowatts or less. The improved fuel efficiency (an improvement of 0.254 gallons per hour) would save 882,579 gallons per month (334 tanker-equivalents) across the fleet of generators.

Recommendations

QoL is an important contributor to logistics burden and soldier perceptions of well-being. Our research has found multiple opportunities to reduce logistics burden while maintaining and, in the case of billets, possibly improving soldier QoL. There is, however, a scarcity of policy and guidance when considering how QoL-related decisions affect the logistics burden. Also, although many promising technologies and material solutions are available to improve water and energy efficiencies, there is no unifying plan for evaluating, fielding, and employing these technologies on base-camp designs.

Because many communities of practice are involved in base-camp QoL determination, we have chosen to organize the recommendations by doctrine, organization, training, materiel, leadership and education, personnel, and facilities. This serves to group the recommendations generally by the Army Staff, secretariat, Army service component command (ASCC), or direct reporting unit that we determine has the lead responsibility in addressing the recommendation.

Doctrine

- Maximize the use of local water sources in support of base-camp operations. Joint doctrine mandates that tactical bulk-water support operations should purify water as close to the user as possible. The Army Corps of Engineers maintains a geospatial database of water sources worldwide. This includes not only lakes and rivers but also groundwater sources and estimated volumes of water. These local sources reduce the need for water convoys and will increase sustainment security as a result. Planners should assess local potable water sources as a selection criterion when conducting base-selection course-of-action analysis.

- Give particular emphasis to integration of water-recovery technology into base-camp operations and to a reassessment of water quality that maximizes the uses for recovered gray water. Base-camp water management does not end after initial consumption of the water, with a significant volume of gray water and black water generated. Typically, wastewater has been simply removed from the base camp via contracted assets, resulting in additional expense and attacks to the gray water–removal convoys. The Army has fielded the SWRS, capable of efficient expeditionary gray-water processing, and continues to develop expeditionary water-recycling and reclamation capabilities. Doctrine should continue to evolve to institutionalize an expeditionary focus and the emerging equipment capabilities.

- Update existing base-camp doctrine to institutionalize the lessons learned since the most current version was published in 2013. Areas that should be addressed include more-detailed QoL management processes, a focus on the "expeditionary" base camp described in multiple Army long-range strategic documents, and the coordination of effort required to deliver QoL at a sustainable level in a way that is consistent with the theater commander's operational priorities.

- Supplement existing documentation with doctrinal information and standards, forming a single base-camp development regulatory standard. Interviews with ASCCs indicate that COCOM regulations on construction and base-camp development are integral source documents in the initial base-camp planning process. These regulations, also known as the Sand Book in U.S. Central Command and known by other colors in other COCOMs, are a rich source of information for the base-camp planner. However, these documents focus on roles and responsibilities, with some discussion of development processes. Nothing in the Sand Book or its sister regulations provides any planning factors or any other means to calculate an order-of-magnitude sustainment-demand estimate. This is a significant shortcoming. Many of these planning factors do exist, but they are found in multiple doctrinal products. The effort to research these planning factors and include them in base-camp development regulations is worthwhile because this should result in better base-camp life-cycle decisions.

Organization

- Consider creating modified table of organization and equipment (MTOE) organizations to construct and administer contingency bases. Doctrinal responsibility for the various aspects of Army base-camp construction, development, and management is distributed across multiple organizations, requiring a coordination of effort that yields mixed results. Other services, particularly the Air Force, have achieved success in standardizing development through the use of dedicated MTOE organizations to deliver contingency bases.

- Assign base-camp management responsibility to an MTOE organization for contingency bases of fewer than 6,000 residents. Regional support groups are doctrinally assigned the commander responsibility for contingency bases with populations above 6,000. However, the vast majority of contingency bases are below that population level. For these smaller contingency bases, units are assigned commander responsibility haphazardly, generally with little or no training on managing base-camp operations. Designating a unit as having the doctrinal responsibility for smaller bases will result in better-trained commanders, resulting in contingency bases that are more effectively and cost-consciously managed.

Training

- Institutionalize base-camp management training. The great majority of base-camp commanders come to that responsibility with little or no specific training for the role. The Army Office of the Assistant Chief of Staff for Installation Management (OACSIM) conducted the only base-camp management training of which we are aware. OACSIM developed a base-camp management curriculum for a 40-hour block of instruction in 2009, based on lessons learned from Iraq and Afghanistan. Although the OACSIM effort was admirable, its effectiveness was limited because it was developed outside the Army's training and doctrine organization. The lesson plans that OACSIM developed should be used as a starting point for the U.S. Army Training and Doctrine Command to develop a base-camp management training program that can be maintained in accordance with existing regulations and practice.
- Develop and institutionalize training and tools that enable COCOM and ASCC planners to more effectively conduct base-camp life-cycle planning. Training would prepare planners to make key decisions about required sustainment levels, based on the availability of food, water, and fuel in the theater and the transaction and opportunity costs that accompany them. Decision-support tools that help planners develop base-camp courses of action should supplement this training. These tools should include cost–benefit analysis modeling capability and a standards-based menu of options that baseline sustainment requirements.

Materiel

- Expand the Army's AMMPS inventory as a replacement for other tactical generators currently in use. AMMPS generators can be linked together to form tactical microgrids, which are demonstrably more efficient than single generators.
- Continue to build capability for base-camp water management. Programs of record exist for current and future solutions for base-camp management of wastewater recovery, treatment, and reuse, as well as potable-water storage and packaging compatible for smaller contingency bases. These solutions all contribute to sustainment security by making contingency bases more self-sufficient for water and therefore minimizing the convoys required to bring bulk water to the contingency bases and remove water.
- Develop and implement minimum energy-efficiency standards for temporary structures. This report discusses at length the impact of shelter energy efficiency on base-camp energy and sustainment security. The Army should establish a minimum energy-efficiency standard for procurement and development of temporary structures that ensures that the benefits of energy efficiency can be sustained into future acquisition cycles.

Policy

- Develop policies that constrain automatic expansion of base-camp facilities and services in support of the combatant commander's operational priorities. Recent history demonstrates that base-camp commanders and tenants arrive at a base camp at the beginning of their rotations and strive to "improve their fighting position" in the name of "taking care of soldiers," as the Army commonly says. This improvement, although well intentioned, often serves to exacerbate the sustainment burden with limited incremental improvement in QoL. Senior leaders should communicate clear guidance to subordinate commanders regarding the acceptable limits for base-camp improvements.
- In tandem with the foregoing, establish—and widely promulgate—standard levels of support for base-camp services and sustainment for each phase of the base-camp life cycle. These levels of support provide a menu-of-options starting point for COCOM planners and base-camp commanders to appropriately sustain a base camp, based on the base-camp life cycle. This provides a baseline by which the base-camp commander can prioritize delivery of sustainment goods and services. The menu of options should be communicated throughout the Army at every opportunity, thereby establishing the new standard for base-camp QoL and managing soldier expectations.
- Establish energy-efficiency requirements or incentives in base-camp energy support contracts. A significant number of contingency bases receive power generation through contracted support. Few of these provide any efficiency or consumption standards that incentivize the contractor to limit its fuel consumption. These contracts should also stipulate that fuel-consumption and energy-efficiency data that the contractor gathers be made available to the Army, unencumbered. Data gaps resulting from contractors' proprietary collection of energy data that were not turned over to us have hampered the research for this report. This same information is essential to the Army for purposes of assessing energy efficiency at individual contingency bases, as well as assessing best practices across a network of contingency bases.
- Eliminate or limit retail restaurants and thereby avoid the water penalty associated with meal preparation and cleanup. As the estimates in Chapter Three indicate, the water penalty can quickly generate additional tankers' worth of water requirements.

Facilities

- When planning, take into account the limitations of minor military-construction (MILCON) thresholds with respect to energy-efficient semi-permanent and permanent base-camp construction. At present, legal thresholds limit the use of operations and maintenance dollars normally to $1 million per project.[1] Beyond this threshold, MILCON dollars are required and must be programmed and are not quickly available during contingencies. The minor MILCON threshold can be insufficient for building generator microgrids to heat and cool facilities in the theater because, if the generators are networked, the sum of the individual units counts toward a single project cost, depending on the legal interpretation of what constitutes a project.
- Develop and implement minimum energy-efficiency standards for semi-permanent and permanent base-camp structures.

---

[1] Some exceptions apply that require secretarial approval and congressional notification or pertain to immediate safety requirements (10 U.S.C. § 2805).

Acknowledgments

The research team benefited significantly from the advice and insights that came from the office of the deputy chief of staff for logistics and from the Army's research and development activities. In particular, we are grateful to LTC Zolten Dukes, LTC Douglas Krawczak, and Desmond T. Keyes, Headquarters, Department of the Army, assistant chief of staff for logistics, operational energy and contingency basing, for their guidance and advice throughout the course of our research. The U.S. Army and Mark A. Leno III of the Logistics Innovation Agency advanced our analysis substantially by sharing their base-camp cost-estimating tool with us.

We also benefited from our interactions with staff members at the U.S. Army Natick Soldier Research, Development and Engineering Center. Our thanks to Claudia J. Quigley and Justine Federici for helping us develop a fuller appreciation of ongoing research and development efforts that target specific base-camp quality-of-life issues and challenges.

At RAND, we thank our program director, Kenneth J. Girardini, for his support and wise counsel.

Finally, we thank our reviewers, Keenan D. Yoho and John Halliday.

Abbreviations

AAFES	Army and Air Force Exchange Service
ACC	Army Contracting Command
AMMPS	Advanced Medium Mobile Power Sources
AOR	area of responsibility
ASCC	Army service component command
ATP	Army techniques publication
ATTP	Army tactics, techniques, and procedures
BCT	brigade combat team
B-hut	barrack hut
BSP	base-supported population
CENTCOM	U.S. Central Command
CHU	containerized housing unit
CLU	containerized living unit
COA	course of action
COCOM	combatant command
COP	combat outpost
CSS	combat service support
DESC	Defense Energy Support Center
DFAC	dining facility
DLA	Defense Logistics Agency
DoD	U.S. Department of Defense
DOTMLPF	doctrine, organization, training, materiel, leadership and education, personnel, and facilities
DX	direct exchange

FMD	Fuels Manager Defense
FOB	forward operating base
G-4	deputy chief of staff for logistics
HEMTT	heavy expanded mobile tactical truck
HQDA	Headquarters, Department of the Army
IMCOM	U.S. Army Installation Management Command
ISO	International Organization for Standardization
kW	kilowatt
kWh	kilowatt-hour
LIA	U.S. Army Logistics Innovation Agency
LOC	line of communication
LOGCAP	logistics civil augmentation program
LSS	latrine, shower, and shave
MILCON	military construction
MOPP	mission-oriented protective posture
morale sat	morale satellite
MRE	meal, ready to eat
MSOW	master schedule of work
MTOE	modified table of organization and equipment
MWR	morale, welfare, and recreation
NATO	North Atlantic Treaty Organization
NSN	national stock number
NSRDEC	U.S. Army Natick Soldier Research, Development and Engineering Center
OACSIM	Office of the Assistant Chief of Staff for Installation Management
OEF	Operation Enduring Freedom
OIF	Operation Iraqi Freedom
OPLAN	operation plan
OPORD	operation order
OPTEMPO	operating tempo
PTSD	posttraumatic stress disorder

PX	post exchange
QoL	quality of life
RC	reserve component
R&R	rest and recreation
SEA	Southeast Asia
SIP	structural insulated panel
SWA	Southwest Asia
SWRS	shower water–reuse system
TEMPER	tent, extendable, modular, personnel
TRADOC	U.S. Army Training and Doctrine Command
TV	television
USAREUR	U.S. Army Europe
USO	United Service Organizations

CHAPTER ONE

Introduction

The Army is committed to the well-being of its soldiers, civilian employees, and their families and has created U.S. Army Installation Management Command (IMCOM) to provide direct oversight of Army efforts toward furthering that goal. The command's mission statement characterizes this commitment: "Our mission is to provide Soldiers, Civilians, and their Families with a quality of life commensurate with the quality of their service" (IMCOM, 2011). Soldiers deployed to contingency bases overseas, such as those in Afghanistan, are not forgotten in this mission. However, unlike permanent and enduring installations, there is no proponent for and limited doctrine when it comes to the management of contingency bases. Instead, it is often left to the base-camp commander to take on the task of continuously improving the foxhole. Toward this end, the Army often goes to great measures to improve quality of life (QoL) on contingency bases.

Many benefits are associated with improved QoL. However, the cost of continued enhancements is less understood and infrequently factored into the decisions regarding QoL on contingency bases. Sustaining QoL in austere theaters with few roads and little infrastructure can be challenging, and prosecuting operations against a determined enemy can add to the challenge. The Army must balance the logistics burden that QoL requirements create and the Army's commitment to provide soldiers with the "right" QoL at contingency bases. That need came to the attention of the Army's deputy chief of staff for logistics (G-4) and the sustainment community and was the subject of this study.

Organization of This Research

We organized the research around four tasks. First, the project described the evolution and importance of QoL enhancements in contingency bases over time. Appendix C traces the evolution of the Army's efforts to provide QoL goods and services from the late 19th century onward and, in particular, the growing institutionalization of those efforts from the 1980s through today.

The second research task was to develop a catalog of the key elements associated with each QoL enhancement and the resources they consumed. The research team analyzed fuel-consumption data on 16 contingency bases in Afghanistan for a period of 39 months. We also analyzed information on the types of generators used in Afghanistan and the typical layouts of contingency bases in theater, including number and types of structures used for billets and number and size of dining facilities and laundry facilities, as well as a database of generators in use in Afghanistan. This information is discussed in Appendix A. We also use historical shipping records to track the flow of QoL-related supplies into Afghanistan. Analysis of these various data sources allowed us to identify the key commodities that underpin most QoL goods and services.

The third task in the research design called for the team to examine the process currently used to plan and execute QoL enhancements. In this regard, the research team conducted a literature review of the regulations and governing directives for QoL provider organizations (e.g., Army and Air Force Exchange Service [AAFES]; morale, welfare, and recreation [MWR]; United Service Organizations [USO]; U.S. Army Corps of Engineers; theater-level base-camp standards). Team members also participated regularly in the Army's community of practice examining base camp–related issues. Finally, members of the team conducted interviews with officers within U.S. Army Sustainment Command (ASC) engaged in QoL management; the Army fellows resident at the RAND Arroyo Center, the Army's federally funded research and development centers for studies and analyses; and officials within the Army Corps of Engineers who design and develop base-camp standards.

The final task in the research design was to develop methods for establishing base-camp QoL packages and their contents. This task drew on insights gathered during the previous three. Because the sponsor for the research was the U.S. Army G-4, the research team concentrated on the G-4's principal instrument for influencing military sustainment operations: making policy.

Categorization of Base-Camp Quality-of-Life Resources and Practices

In this research, we partition the components that contribute to the QoL on a base camp into two categories: (1) personnel services and (2) material services. We view the first category as the things that are provided to enhance QoL and the second category as the means by which those things are provided. In determining the level of QoL provided on the base, commanders can adjust either the things that are provided or the means by which they are provided. Each decision has an implication for the logistics burden created.

Personnel Services

Personnel services include MWR provisions, types of food and water delivered, opportunities for rest and recreation (R&R), religious reflection, and education. The evolution of services available on a base camp is often a function of time. According to Field Manual 1-0 (Headquarters, Department of the Army [HQDA], 2010b), units are expected to use their unit-level recreation kits and athletics and recreation support for the first 30 days of deployment (C + 30). Between C + 30 and C + 60 days, units are expected to coordinate with their chains of command, base-camp commanders, and G-4 for deployment of MWR (HQDA, 2010c) service-level kits. Also in this time frame, policy on rest-area use should be established and measures put in place for distribution of health and comfort packs and class I supplies (food, rations, and water). This is also the time when AAFES is expected to establish a presence (e.g., direct operational exchange—tactical, tactical field exchange, or AAFES imprest fund activity). Between C + 60 and C + 120, depending on circumstances, the Army service component command (ASCC) G-4 will establish theater activity address codes for MWR. Corps and divisions will provide rest areas for brigade-sized units, and the component manpower or personnel staff officer can establish pass programs and implement R&R policies. This is also the time frame in which, theater conditions permitting, the USO and Armed Forces Entertainment begin operations.

Army Regulation 215-1 (HQDA, 2010c) governs military MWR programs and nonappropriated-fund instrumentalities. The regulation anticipates that units about to deploy will ship a 30-day supply of athletic and recreational equipment and a library book kit and states that the commander of IMCOM will "[s]upport MWR requirements for deployment/mobilization and contingency operations, to include designation of emergency essential (E-E) civilians, as appropriate" (p. 3, ¶2-3[u]). The regulation also provides units with the means to operate athletic activities and recreation programs, unit lounges, and similar activities supported by AAFES imprest funds. Units can also draw on Common Table of Allowances 50-909 (U.S. Army, 1993) for recreation and athletic kits. Larger units, including corps, division, and brigades, have access to MWR service-level kits, videocassette recorders, televisions (TVs), computers with game packs, weights, karaoke, and keyboard instruments. Army Regulation 215-1 also provides for entertainment from the USO and Armed Forces Entertainment, military clubs, R&R centers, unit lounges, and rest areas in the theater or corps area.

Material Services

Material services are defined by the physical infrastructure of the camp, such as types and number of billets, dining facilities (DFACs), laundry, toilets and showers, and gyms and gym equipment.[1] The engineering standards for camps define the physical environment for delivery of QoL. The two biggest determinants of camp characteristics, facilities, and services (and therefore QoL) are the size of the camp's population and the anticipated duration of operations. These standards account for the full suite of services and facilities, including gyms, MWR, chapels, theaters, and education centers. They reflect a time line and establish facility standards along it: Initial camps for periods of less than six months, temporary camps for periods of operation of six to 24 months, and semi-permanent for camps with a life expectancy of two to 25 years (see U.S. Army Europe [USAREUR] and Seventh Army, 2004, Table 4.1). For each time frame indicated, there are distinct standards for defining the DFAC, housing, latrines and septic system, showers, and other facilities at a base camp. The population size is also a factor: Generally, the larger the population, the greater the abundance of facilities and services.

In addition, some directives provide three different levels of development for contingency bases: basic, expanded, and enhanced (HQDA, 2013a). **Basic** capabilities "are established as part of initial entry and are implemented primarily using organic capabilities and prepositioned stocks" (HQDA, 2013a, p. 1-2). **Expanded** capabilities "are basic capabilities that have been improved to increase efficiencies in the provision of base camp support and services, and expanded to sustain operations for a minimum of 180 days" (HQDA, 2013a, p. 1-3). **Enhanced** capabilities

> are expanded capabilities that have been improved to operate at optimal efficiency and support operations for an unspecified duration. These capabilities are flexible, durable, and near self-sustaining, and implemented primarily through contracted support. Many of the functions, facilities, and services and support resemble those of a permanent base or installation. (HQDA, 2013a, p. 1-3)

---

[1] DFACs are used only at "large-enough" installations. DFACs are popular because deployed personnel highly value food quality.

Organization of This Report

Chapter Two focuses on defining QoL on a base camp, Chapter Three examines opportunities for savings, and Chapter Four provides conclusions and recommendations. Appendixes that capture the details of our research support the analysis presented in these chapters:

- Appendix A provides a review of a sample of base camps in Afghanistan. We discuss the services available on each camp and the demand for fuel and other support required to sustain it.
- Appendix B describes how we used crowdsourcing to gather opinions and experiences from soldiers who were deployed to and spent time in base camps.
- Appendix C presents research on the importance of base camps to soldiers' physical and psychological well-being.
- Appendix D describes the energy efficiency and other physical characteristics of shelters and how those affect fuel consumption.

What Is Quality of Life?

An adage in the military, as well as the world at large, states, "when you don't know where you're going, any road will get you there." So it is with QoL. Although the term is generally understood within a popular context, it means different things within different Army and U.S. Department of Defense (DoD) communities. The Army's definition of *QoL* on a base camp might conflict with the great majority of these definitions having little applicability to life in a potentially austere operating environment in which there is little or no infrastructure. A prudent start to this discussion is to establish a framework to discuss QoL in light of the unique mission and security constraints placed on contingency bases.

Before that discussion, however, we take a short detour into terminology. The term *base camp* is often used to describe the concept of a contingency base. In fact, the basic Army doctrine for these concepts is titled *Base Camps*. The DoD equivalent term is *contingency base*, and the two terms describe essentially the same concept. Because these locations are routinely occupied by multiple services, we have chosen the term *contingency base* as the term more appropriate for joint operations. For this report, we define *contingency base* by the doctrinal definition of *contingency location*: "[a] non-enduring location outside of the United States that supports and sustains operations during named and unnamed contingencies or other operations as directed by appropriate authority" (Joint Chiefs of Staff, 2018, p. 51). This is in contrast to installations, which are permanent military facilities within the continental United States, and enduring facilities, which are semi-permanent facilities outside the United States.

In a general context, *QoL* is defined from the perspective of the individual. For instance, *Definitionsfor* defines *QoL* as one's "personal satisfaction (or dissatisfaction) with the cultural or intellectual conditions under which [one] lives (as distinct from material comfort)" ("Quality of Life," undated). Efforts to define the term date back to at least the 1970s, and usage of the term goes back further than that. Numerous QoL definitions, independent of military usage, attempt to summarize personal satisfaction on cultural, political, and economic grounds. All of these definitions focus on a central theme—in society, *QoL* is defined and experienced by individuals. Each of us defines *QoL* in terms of our own desires, motivations, and aspirations. By contrast, base-camp QoL must be defined collectively for the entire base-camp population, almost always by the base-camp commander on behalf of the population.

The QoL levels of support that are provided at a base camp can be based on several factors, including the type of mission and its expected duration, the security situation, the maturity of the base-camp infrastructure, and the resiliency of the lines of communication. The base-camp commander is in the best position to assess all of these factors to achieve the best balance of mission and QoL levels; in fact, joint doctrine (HQDA, 2010b) charges this commander as the responsible party for determining base-camp QoL support levels.

This being the case, the base-camp population must subordinate individual QoL expectations to conform to the operational realities imposed on the base-camp commander. However, there are excellent reasons for the commander to consider QoL desires of the base-camp population and to meet those when possible—in part, because considering the soldiers' preference will help camp commanders balance the beneficial effects of QoL against the sustainment burden they create.

Soldier Preference and Value of Quality-of-Life Features on Contingency Bases

In this section, we rely on several sources of information to draw conclusions about how soldiers value different QoL elements and how QoL influences soldier readiness. Although a direct connection between QoL and readiness has not been established, multiple studies examine the role of QoL on both the physical and psychological well-being of soldiers and on their ability to withstand and recover from stress factors (see Appendix D). Many of the psychological factors described in these studies fall outside the scope of this research in that they refer to an individual's intrinsic ability to manage stress and trauma. Most studies establish domains or distinct areas that influence soldier well-being, such as housing, family support, MWR opportunities, physical fitness, and unit camaraderie.

Extensive research on stressors affecting soldiers deployed during operations Desert Shield and Desert Storm is chronicled in the 2000 RAND National Defense Research Institute publication *Psychological and Psychosocial Consequences of Combat and Deployment with Special Emphasis on the Gulf War* (Marlowe, 2001), which we summarize below and in greater detail in Appendix D of this report. That study focused on combat arms units rather than including support groups because the authors determined that these were at higher risk for psychological and psychiatric trauma than support groups (Marlowe, 2001).

Several notable infrastructure- and capital-based stressors adversely affected troops. Most of these stressors are unique to temporary bases. The nature of the deployment, in which soldiers were isolated from the host nation almost entirely and temporary bases were constructed away from cities and the amenities that accompany them, left soldiers without chances to interact with host cultures or the population.

Furthermore, bases seemed to be hastily built, overcrowded, and without much room for privacy or personal time. Marlowe heavily emphasized issues with living space.

A lack of MWR equipment compounded the effects of isolation. Not only were troops unable to get away from each other; they could not use their time to relax with each other.

Issues with living space, a lack of interaction with local culture, the general location of theater (the Arabian Desert), and an additional lack of capital to help offset these issues on base (MWR equipment) provided high levels of stress and drove a sense of "cabin fever" among troops.

Other points of infrastructure deficiencies contributed to reduced morale and presented additional stressors. A lack of reliable mail service and telephone communication infrastructure contributed heavily to unease and increased uncertainty surrounding soldiers' personal lives back home.

Organizational requirements also acted as stressors. Long days of work (14 or 15 hours per day for six or seven days a week in many cases), sleep deprivation, and severe physical stress contributed to the overall sense of stress. The use of "fillers," or personnel who were rapidly deployed to fill voids in units and who lacked cohesion and rapport with current unit members, contributed to a sense of confusion, increasing stress. Additionally, troops in some organizations were still eating meals, ready to eat (MREs)—Army emergency combat rations—months after arriving in the field (Marlowe, 2001).

Many of Marlowe's findings are echoed in a small crowdsourcing survey that RAND researchers conducted in 2014 for the current project, detailed in Appendix B. The top contributors to QoL that respondents listed were hot meals, laundry service, communication home or internet, gym, and showers or running water. The top detractors of soldier QoL mentioned in the crowdsourcing survey were lack of sleep, unhygienic conditions, extreme temperatures, and poor leadership. The top desirable elements listed were routine Skype and free internet, quiet and privacy, and better sleeping conditions.[1]

The U.S. Army Natick Soldier Research, Development and Engineering Center (NSRDEC) recently completed a survey of 1,200 soldiers at four locations (Federici and Augustyn, 2015). The purpose of the survey was to enable leaders to make informed decisions about the balance between QoL services and resource constraints by eliciting the "voice of the soldier," as the Army commonly calls it, in identifying how they prioritize QoL-related base-camp services. The survey asked soldiers to express how billets, field feeding, MWR, hygiene, work area, personal security, and spiritual and psychological support contribute to their perceptions of QoL. Preliminary results from the survey indicate that soldiers weight billets and field feeding almost equally in terms of their ability to increase QoL, followed by MWR,

Figure 2.1
U.S. Army Natick Soldier Research, Development and Engineering Center Preliminary Soldier Quality-of-Life Survey Results

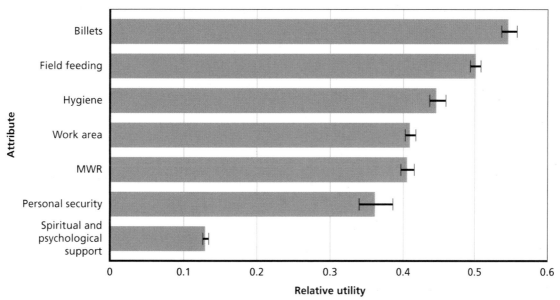

SOURCE: Federici and Augustyn, 2015.
RAND RR1298-2.1

---

[1] "Nothing" was also listed as a top option.

hygiene, and work area. Personal security and spiritual and psychological support ranked least influential on QoL.

The survey also asked soldiers to provide a "swing" score for each attribute to characterize the increase in QoL that improving camp conditions can achieve, from the lowest to highest levels. The results indicate that improvements in comfort (type of bed and climate control) rank highest, followed by access to unclassified computers, availability of supplements and enhancement food items, and shower frequency.

Defining and Prioritizing Quality of Life for Contingency Basing

The preceding discussion establishes billeting (including privacy), food, communications home (including access to Wi-Fi and Skype), and hygiene as the QoL elements important to soldiers. By doctrine, the base-camp commander is the person responsible for the base-camp QoL. However, we still need to define precisely what the commander is responsible for. Army doctrine and policy are less than clear when defining *QoL*. At one end of the spectrum, U.S. Army Techniques Publication (ATP) 3-34.40 states, "The quality of life captures the operational commander's requirements for bed-down and base camp living standards" (HQDA, 2008, p. E-1), which is too narrow a definition to be useful to the commander. At the other end, Navy studies expanded the notion of QoL to embrace the entirety of Maslow's hierarchy of needs (Kerce, 1992). To effectively scope the commander's base-camp QoL requirements, the definition must apply directly to the expeditionary environment.

ATP 3-37.10, *Base Camps*, provides a reasonable starting point (HQDA, 2013a). This document serves as Army doctrine for base-camp planning and execution. In the discussion on base-camp capabilities, ATP 3-37.10 describes levels of base-camp capabilities as

> the characteristics of a base camp in terms of support and services (overall quality of life [QoL]) that are provided and the nature of the construction effort applied that are commensurate with the anticipated duration of the mission. (p. 1-2)

Although it is not a definition of *QoL*, per se, the preceding statement implies two critical elements essential to the definition. The first is the acknowledgment of a responsibility to effectively manage base-camp support and services. The second is that QoL is directly related to the duration of the mission. Indeed, ATP 3-37.10 defines *levels of capabilities* as directly related to the base-camp life cycle, with capabilities increasing as the base camp matures (HQDA, 2013a).

Although ATP 3-37.10 is specific on the base-camp life cycle, it does not define what support and services constitute QoL. However, *Commander and Staff Officer Guide*, ATTP 5-0.1 (HQDA, 2011), provides logical categories for QoL support and services. Annex F of ATTP 5-0.1 is the format for the sustainment annex of the base operations order. Within Annex F, the sustainment paragraph consists of four major subparagraphs:

- material and services: classes of supply, field services, transportation and distribution, maintenance, contract support, and contract labor
- personnel: morale development and support, headquarters management, and strength maintenance

- health system support: preventive medicine, medical and dental services, and veterinary services
- host-nation support as required.

These subparagraphs capture the vast majority of QoL-related commodities and services. In particular, the commanders' management of material and services and personnel will create the largest QoL impact on the base camp. Table 2.1 outlines examples of QoL elements in the material and services and personnel categories.

There are interesting parallels between the QoL elements in Table 2.1 and Maslow's hierarchy of needs.[2] The material and service elements tend to correspond to Maslow's physiological and security needs. In this instance, physiological needs are addressed through housing and food and through hygiene services provided to address the disease vectors that endanger security. As we move to the personnel elements, we see the progression to Maslow's higher-level needs, including self-actualization needs that spiritual and religious support represents. That the QoL elements listed span the entire range of Maslow's needs illustrates the importance of managing these elements effectively. To a great extent, striking the correct balance between mission accomplishments and providing the level of QoL to the base-camp population that helps ensure mission readiness will define success for the base-camp commander.

What Requirements Does Quality of Life Produce for the Sustainment Community?

One of the key tasks for the sustainment community is to support the force and its ongoing operations. The nature of this support varies greatly, depending on the circumstances that the operation presents. The threat, the quality, the maturity and abundance of lines of communication (LOCs), the maturity of local infrastructure, the geography of contingency bases, and the availability of contractors are but a few of the factors that affect sustainment. Many of these factors increase the challenges posed to the sustainment community and increase the need to prioritize essential goods and services. The sustainment community will, under such circumstances, need sound policies that set priorities for the delivery of QoL, given the conditions in the theater. The remainder of this section summarizes QoL-related sustainment requirements.

Table 2.1
Quality-of-Life Elements, by Category

Category	Element
Material and services	Housing, meals and feeding, shower, latrine, and waste removal and recycling
Personnel	Chapel and religious services, internet café, fitness center, AAFES, movie theater, telephone call center, educational center, other vendor services (including local bazaar, post office, and banking), USO, and base-camp transit or bus route

---

[2] Maslow's hierarchy of needs is often depicted as a pyramid; the foundation of the pyramid is physiological needs (food, water, and sleep), and the pyramid builds up toward higher-order needs. Maslow postulated that successively higher needs are not achieved if basic needs are denied. The order of needs, from low to high, is physiological, safety, love and belonging, self-esteem, and self-actualization (Maslow, 1943).

Quality-of-Life Requirements, by Phase of Military Operations

Joint doctrine recognizes six phases of military operations, as shown in Figure 2.2. Phase 0 (shape) missions are designed to dissuade or deter adversaries and assure friends, as well as set conditions for the contingency plan. These activities are generally conducted through security cooperation activities, such as building partner capacity and multinational military exercises. The majority of these events tend to be of short duration, often rotational, and operate from a partner-provided, sometimes U.S.-enhanced facility. Examples include Camp General Basilio Navarro in the Philippines; Cairo West Air Base, the former site of Operation Bright Star; and Multinational Force and Observer camps in support of the Egyptian–Israeli Treaty of Peace. They make relatively small demands on the sustainment community for QoL, which can be turned on and off with the deployment cycle of the troops involved.

Phase I (deter) QoL requirements can vary substantially, depending on the duration of the deterrent operations. In this phase, adversaries are deterred from undesirable actions because of friendly capabilities and the will to use them. Deterrence is generally weighted toward security activities that are characterized by preparatory actions to protect friendly forces and indicate the intent to execute subsequent phases of the planned operation. In short-duration shows of force, such as Operation Desert Shield in 1990, the conditions can be austere and the demand for QoL goods and services confined to occasional hot meals, two showers per soldier per week, 15 pounds of laundry, and rudimentary quarters and sanitation. The demand for QoL

Figure 2.2
Joint Operation Phases Versus Levels of Military Effort

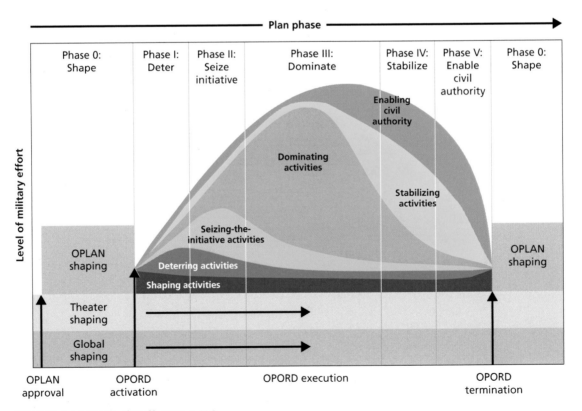

SOURCE: Joint Chiefs of Staff, 2011, p. V-6.
NOTE: OPLAN = operation plan. OPORD = operation order.
RAND RR1298-2.2

can increase quickly as Phase I grows longer, as the case of Kuwait in 2003 illustrates. In that instance, Kuwait grew from one to 14 camps between January and July; by August, the Camp Wolf DFAC (the camp was located on the international airport grounds as the major reception, staging, onward movement, and integration site) was serving 7,500 meals four times daily and operating an air-conditioned AAFES store (Fontenot, Degen, and Tohn, 2004). Because deterrence operations typically take place on friendly soil (e.g., Saudi Arabia, the Philippines, Kuwait, West Germany), the local economy often contributes a good deal to QoL through retail and contracted goods and services. Long-term deterrent postures, such as that of USAREUR between 1949 and 1991, should tend to produce markets and providers that cater to U.S. troop needs.

Phases II and III (seize the initiative and dominate) seek to seize the initiative in all situations through decisive use of joint force capabilities and seize the enemy's will to resist or, in noncombat situations, to control the operational environment. Recent Phase II and III operations have seen QoL goods and services significantly curtailed. Most soldiers in Operation Iraqi Freedom (OIF) received no mail from the time their units crossed the border into Iraq until the seizure of Baghdad. Although OIF might shape the modern soldier's expectations for the duration of this phase, examples from World War II remind us that these phases can be long (e.g., Operation Torch in North Africa was seven months, the Italian campaign 11 months, and the liberation of France nearly three months) and that, in such cases, non-subsistence goods and services (e.g., hot meals, showers, mail, communications home, and in-theater R&R) might not be essential to sustain the fight.

Phase IV (stabilize) is typically characterized by a shift in focus from sustained combat operations to stability operations. OIF and Operation Enduring Freedom (OEF) required a protracted Phase IV operation, which resulted in an ever-larger commitment from the sustainment community to support QoL demand. Some of this expansion reflects the providers' determination to take good care of the troops. The engineering community does this through incremental improvements in living quarters and conditions, the MWR community by expanding its recreation and entertainment menus, and AAFES by deploying a larger footprint in the theater and offering more goods and services through its retail operations and concessionaires. The sustainment community underwrites all of this by shipping the commodities supporting this, including the fuel that drives the improvements and sustains the base-camp QoL. The next section of this report describes QoL-related requirements that fall to the sustainment community to satisfy.

Major Commodities

Major commodities are the principal source of demand for base-camp sustainment support. Their relative abundance at contingency bases is also a key indicator of relative QoL. The demand for these commodities moves linearly with the size of troop populations and their preferences, as described in this section.

Food

Food (supply class 1) has consistently accounted for 40 percent to 65 percent of second-destination transportation tonnage delivered between 2008 and 2013 in Afghanistan (see Figure 2.3). Sustainment community–run ground convoys usually deliver rations destined for DFACs.

Figure 2.3
Second-Destination Transportation Transactions from August 2010

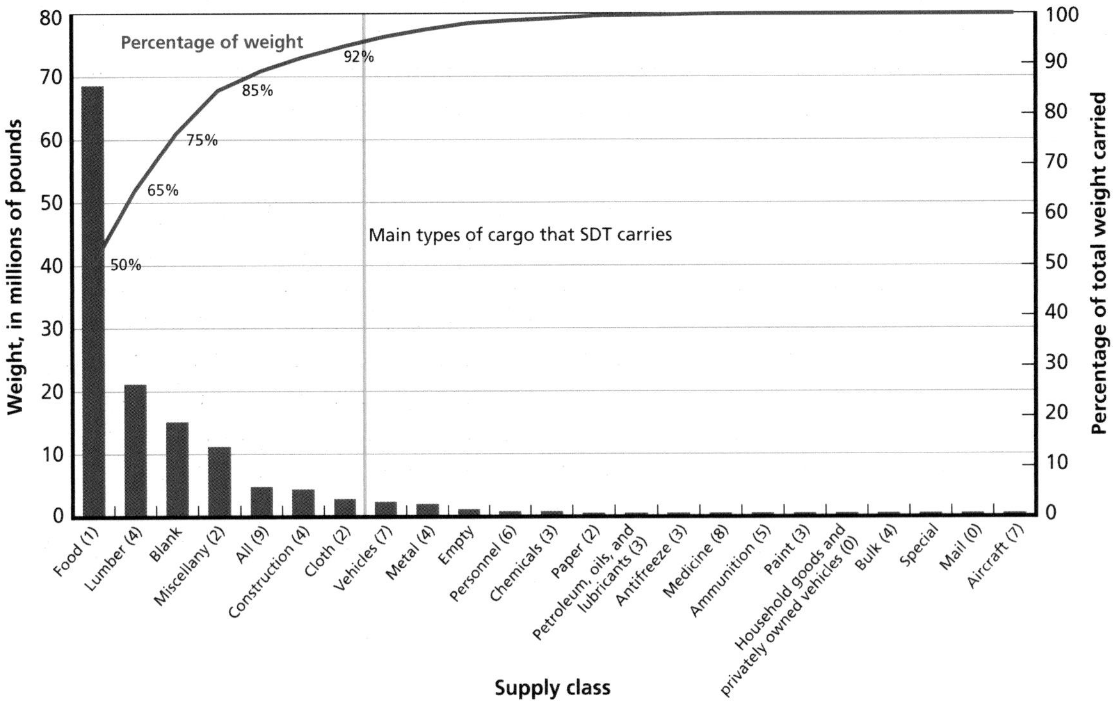

SOURCE: Second-destination transportation transactions for August 2010.
NOTE: In addition to the supply classes shown, these transactions moved ammunition, fuel, and water.
RAND RR1298-2.3

The fresher the food is, the more it weighs. The research team hypothesized a division-level task force of five brigade combat teams (BCTs) and their enablers, or 19,283 personnel, then estimated their food requirements for a year (21.1 million meals). If fed prepackaged MREs, the force would require 12,220 tons of these rations (1,111 M977 heavy expanded mobile tactical truck [HEMTT] loads). A B-ration (meals prepared using canned or preserved ingredients) diet would weigh 17,596 tons (1,600 M977 loads). An A-ration (meals prepared using fresh, refrigerated, or frozen foods) diet for the force would weigh 19,283 tons (1,753 M977 loads).[3]

In reality, DFACs at large forward operating bases (FOBs) feed fresh food, often four meals per day. Packaged rations tend to be consumed during operations and while manning combat outposts (COPs). Consumption and spoilage drive replenishment rates. Members of 377th Theater Sustainment Command interviewed for an earlier project reported that all of the convoys they operated in Iraq involved delivery of rations. Of course, soldiers must always be fed, no matter where they are stationed, so food deliveries are unavoidable. Only their frequency is negotiable, and then only within the time frame of the shelf life of the foodstuffs required.

Water

Providing water is a major task for the sustainment community. Arid and desert conditions, such as those encountered in Afghanistan and Iraq, add to the degree of difficulty. The "inte-

---

[3] Calculations based on data found in NSRDEC, 2012.

grated planning factor" for potable water in arid climates is 15.54 gallons per person per day (U.S. Army Combined Arms Support Command, 2008). For a force the size of the U.S. Army in Afghanistan in 2013 (54,370) (U.S. Senate Committee on Appropriations, 2013), the annual requirement would be 308.8 million gallons. Other sources, however, suggest significantly higher consumption rates: 60 gallons per person per day, or approximately 1.2 billion gallons for the Army force in Afghanistan in 2013 (USAREUR, undated).[4]

Fuel

All base-camp energy sources depend on fuel. Fuel is needed to run the camp generators, which, in turn, provide electricity for heating and cooling, lights, computers, dining, medical maintenance, and all other facilities. A review of the allocation of fuel between operations (e.g., flights, fueling of equipment) and contingency base operations at 16 contingency bases in Afghanistan indicates that the majority of the fuel consumed went to support contingency base operations (the only deviation from this finding was at large airfields, such as Bagram and Kandahar). As an illustration, Figure 2.4 shows power requirements for Spin Boldak, a contingency base in southern Afghanistan that had primarily an Army population throughout its lifespan. According to empirical data for this contingency base, billeting was the largest user of power generation from July 2012 to February 2013, requiring 61 percent of the power gener-

Figure 2.4
Power-Generation Requirements, by Facility Type on Spin Boldak, July 2012 to February 2013

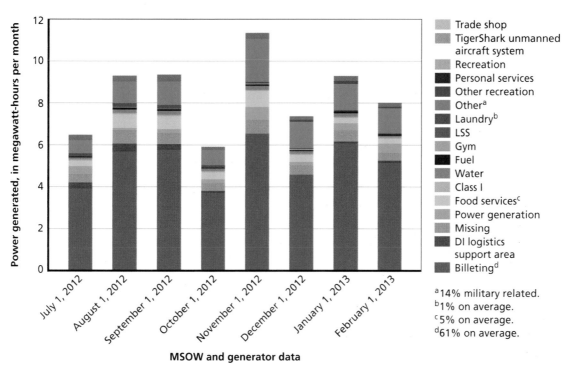

[a]14% military related.
[b]1% on average.
[c]5% on average.
[d]61% on average.

SOURCE: Logistics civil augmentation program (LOGCAP) data.
NOTE: MSOW = master schedule of work. LSS = latrine, shower, and shave.
RAND RR1298-2.4

---

[4] This level of consumption might reflect DFAC, motor-pool flight line, hospital, laundry, sanitation, and morgue requirements presented on a per capita basis.

ated. Power generation is directly related to fuel because the power is supplied solely through generators. Military-related activities, such as maintenance shelters, tactical operating centers, or military administration facilities, made up the second-largest category of power requirements, requiring, on average, 15 percent of a base's power requirement, and food services made up the third largest, at 5 percent of the power requirement. Typical MWR-related facilities, such as gyms and personal service–related structures for recreation, and AAFES required less fuel.

Morale, Welfare, and Recreation

MWR brings soldiers recreation (e.g., movies, television, internet, voice over internet protocol, games) and an opportunity to "blow off steam" from the day's events. Base-camp plans provide space for MWR facilities. Square footage and building permanence are allocated according to the size of the population the program must support. The MWR program has its own funding. In the last year that it received overseas contingency operation funds, fiscal year 2011, the program received $9,280,000. Overseas contingency operations aside, MWR-appropriated funds, which generally support overseas operations, allow us to estimate the per capita spending on MWR.

Given that the MWR space is part of the base-camp engineering standards and thus represents an unavoidable construction cost and that the program operates on its own funding, the MWR requirements that fall to the sustainment community are limited. Army logisticians must provide fuel to generate electricity for these requirements and must provide life support for the MWR professional staff (e.g., food, water, and quarters). MWR staff operate and maintain the facilities. Because MWR staffing levels are minimal, their impact on food, water, and fuel consumption are unlikely to be significant at the contingency base level.

Army and Air Force Exchange Service

Much like MWR, AAFES has its own funding (largely from its retail operations) and personnel. It delivers its support in packages suitable to the circumstances, from a field exchange and "post exchange (PX) in a box" to full retail stores. Its facilities are a planned part of the camp footprint for all larger contingency bases. In 2008, AAFES operated 90 exchanges and restaurants in Iraq and another 224 in Afghanistan and the wider Middle East (AAFES, 2009). AAFES facilities depend on the host contingency base and thus the sustainment community for fuel, energy, and water. We address the demand for food, water, fuel, and energy in the next section.

Summing Up Quality of Life–Related Demand on the Sustainment Community

It appears from the description of demand presented in this part of the report that QoL-related demands on the sustainment community arise principally, though not exclusively, in the key commodities of food, water, fuel, and energy. MWR, AAFES, and AAFES concessionaires do not produce significant demand signals on their own outside of these commodity classes. The demand on the sustainment community from food extends only to its obligation to feed Army personnel in the theater of operations. The demand signal itself is unavoidable, but the transaction costs that accompany demand are subject to management (e.g., change the types of rations to be delivered and the modes of delivery), discussed in later sections of this report. Water and fuel appear to be the most challenging commodities to sustain (because they are both bulky and heavy and require specialized trucks to deliver them) and might have the greatest impact on QoL. The abundance of local water sources figures directly in how frequently camp residents can shower, the quality and availability of laundry service, the cleanliness and aesthetics surrounding shower and latrine facilities, and the availability of retail restaurant meals. Fuel is the prime mover for energy at the camps and is therefore key to many MWR goods and services—most importantly, communications with home. Fuel ultimately drives climate control, which is fundamental to numerous QoL-related factors. Likewise, fuel makes personal electronics possible by providing electricity to recharge batteries and drive small appliances, although the energy demand from personal electronics is an exceedingly small portion of the total contingency base consumption.

CHAPTER THREE

Opportunities for Improvement

This chapter considers how base-camp commanders should balance the QoL services provided to a base-camp population with the new realities of budget and mission. The levels of service provided and soldiers' expectations vary by phase of operation and duration. The duration of an operation and how long a base camp will exist are often difficult to predict, limiting opportunities for long-term strategic planning. The evolution of QoL support on a contingency base means that facilities and infrastructure are sometimes designed for expediency, with limited information or understanding of the implications that these decisions can have on long-term sustainment.

Benefits to Effective Contingency Base Management

There are several reasons base-camp commanders should aspire to optimizing contingency base effectiveness:

- personnel security: Since 2001, the Army suffered more than 2,000 soldier casualties conducting ground resupply convoys, which consist of 70- to 80-percent fuel and water by weight. By adopting energy-efficient strategies that might include more-efficient generators and microgrids, renewable energy technologies, and effective waste-recovery systems, base-camp commanders can reduce the logistics footprint. This reduces resupply convoys, directly resulting in improved force protection for U.S. soldiers.
- energy security: The Army Energy Security Implementation Strategy notes that 59 percent of energy consumed during contingency operations goes to generators and facilities. The base-camp commander controls these directly. Reducing energy consumption will reduce operational risk to contingency bases by lowering their dependence on a critical resource that is a prime target for enemy interdiction.
- cost efficiency: In an era of more-constrained resources, commanders will have to be even more-effective stewards of Army resources to ensure that contingency bases function effectively. Increasingly, cost–benefit analysis will become part of course-of-action (COA) analysis.

Policies Versus Standards

Army standards exist that influence QoL elements, but nothing currently documents Army-wide directives for how the base-camp commander fosters QoL.[1] These standards exist in several forms. Numerous Unified Facilities Criteria outline engineering guidelines that govern base-camp construction. Army regulations and doctrine establish standards for QoL services, generally expressed as "rule-of-thumb" ratios. These standards tend to determine what is provided given a population of soldiers. For example, at a base camp supporting a population of 1,000 personnel, the standards prescribe a certain number of toilet seats and showerheads, a certain amount of square footage for sleeping quarters specified by the individual's grade, a chapel, a DFAC, and so on. Commodity-management standards will dictate some number of showers per person per week and some level of laundry service (e.g., 15 pounds per person per week). These types of standards do not address other important aspects of QoL management, especially the questions of how much is enough, the conditions under which this QoL can be delivered, and the length of time for which it can be sustained.

The prevailing engineering and service standards also include time-phased improvements. Contingency bases are expected to evolve through their life cycles from a basic level of service to an expanded and, eventually, an enhanced level with the passage of time.[2] A deterministic effect results from time-phasing the evolution of contingency bases and their corresponding services. This determinism could undermine local commanders' ability to control these processes and subordinate them to his or her command priorities and imperatives—or contain the demand they generate for sustainment-community support.

At present, there is a general lack of policy to establish how much QoL support is required, given the circumstances in theater, the phase of military operations under way, the robustness of the theater distribution system, or the level of permissiveness (i.e., threat level and relative freedom of action enjoyed by U.S. forces) in the area of responsibility (AOR). The closest approximation is the combatant-command (COCOM) base-camp development standards for an AOR, exemplified by the U.S. Central Command (CENTCOM) Sand Book (CENTCOM, 2009). The Sand Book consolidates applicable doctrinal responsibilities and rule-of-thumb planning factors into a single document and adapts these standards to the specific needs of an AOR. The Sand Book, and similar documents developed by other COCOMs, provides the best starting point for developing Army QoL policy.

Command, Control, and Administration

The process of identifying contingency bases and their commanders begins at the geographic COCOMs. The commands assess the operational environment to determine basing require-

---

[1]   The view presented here is consistent with the results and recommendations from the U.S. Army Training and Doctrine Command (TRADOC) and Army doctrine, organization, training, materiel, leadership and education, personnel, and facilities (DOTMLPF) (TRADOC, 2013).

   Other DOTMLPF Integrated Capabilities Recommendations examined base-camp management, base-camp standards, and base-camp sustainment. Each report captures results from TRADOC capability-based assessment and the gaps the analysis identifies.

[2]   See USAREUR and Seventh Army, 2004, Table 4.1, for evolution of engineering standards. See HQDA, 2010b, for the evolution of MWR standards.

ments. The combatant commander establishes, in coordination with the involved DoD components, contingency-basing criteria in OPLANs and supporting plans. Those criteria result in specific base-camp locations developed through COCOM staff coordination, including the input of strategic planners, logisticians, and engineers.

The combatant commander also tasks the lead service component with the responsibility to manage that base camp. Typically, the commander makes that tasking based on the preponderance of personnel planned for that location, but other operational considerations can take precedence. For example, a base camp with an improved airfield might have a population of mostly Army personnel. However, the specialized requirements of operating and maintaining an airfield might necessitate the Air Force being assigned as the lead service for that base camp.

Command and control and administrative operation of contingency bases are an evolving process. TRADOC Pamphlet 525-7-7 (TRADOC, 2009) noted that many key questions about base-camp command and control and administration have not yet been answered definitively. As a result, the exact nature of relationships between units tasked to operate contingency bases and their tenants are often ad hoc.[3] Regional support groups are doctrinally assigned to manage any contingency base with a population of 6,000 or above (including soldiers, civilians, and contractors). However, no doctrine or policy exists that tasks a particular modified table of organization and equipment (MTOE) unit with the mission for camps below the 6,000-population threshold. This is a significant gap because the vast majority of contingency bases have fewer than 6,000 residents. Camps below this level wind up with base-camp commanders having no more than 40 hours of training specific to the base-management function.

Nevertheless, the spectrum of camp-management options can be understood. At smaller contingency bases, tenant units also run the camp. Larger bases utilize a "mayoral cell," described by TRADOC Pamphlet 525-7-7 as the "organizational staff and structure in charge of the internally focused operations and administration of a single base camp, including the related aspects of master planning, construction, operations and management, facilities maintenance, security, and sustainment" (p. 9, ¶2-3[d]). Satisfying these responsibilities might also involve a provost marshal and military-police platoon or company, a base defense operations center, and additional personnel to supervise the life-support commodity areas, e.g., billeting, contingency base energy, DFAC operations, sanitation, pest and vector control, and waste management. In addition, the unit administering a base camp must maintain liaison with the tenants, including AAFES, MWR, and contractors.

Commanders are generally instructed to deploy QoL resources as mission, enemy, terrain and weather, troops and support available–time available, civil considerations, and resources allow (HQDA, 2010b). However, recent conflicts in Iraq and Afghanistan have accustomed the Army to permissive sustainment environments; rare was the circumstance in which sustainment was interrupted for a significant period of time as the result of enemy activity. The more-serious disruptions of sustainment often resulted from political considerations having nothing to do with enemy activity. This level of relative mobility cannot be used as the planning standard for future conflict. Indeed, the concept of anti-access and area denial anticipates this, envisioning future conflicts of semi-permissive or non-permissive environments. In this scenario, the mobility of sustainment convoys cannot be ensured. In particular, fuel convoys can reasonably be expected to be particular targets of enemy interdiction. To a degree not

---

[3] See TRADOC, 2012, for an analysis of base-camp management gaps.

experienced in recent memory, this scenario will require better management of base-camp resources to weather the inevitable supply interruptions.

As subsequent sections of this report describe, a lack of clear policies establishing a basis of issue for material and services, personnel elements, and other QoL components can generate additional, unanticipated (and potentially unsupportable) demands for fuel, electricity, and water—among other things.

So how does the Army execute the base-camp mission under anti-access and area denial? The answer lies in a DOTMLPF analysis of the problem and committing to a combination of short-term and long-term efforts that make the base camp more resilient to sustainment interruptions. This chapter focuses on sustainment of the key commodities: food, water, and fuel.

Improvements, by Commodity Area

Food

Napoleon Bonaparte is attributed with the military maxim, "an army marches on its stomach." This was not hyperbole on Napoleon's part: He lost more soldiers to spoiled food than to battle. We know, as Napoleon understood then, that anyone who has ever served appreciates food as an important morale factor. The research team encountered anecdotes throughout the project about complaints at camps where rumors suggested that the variety of foods available might be reduced. Other anecdotes described sagging morale when the fourth meal, "midrats," was rumored to be ending. Because of the security and cost issues discussed previously, four meals of fresh food might not be realistic on future battlefields. Some balance between fresh food and prepackaged Army meals must be considered. DFAC meals do not appear to be a rich source of savings for the sustainment community.

Our research finds that, as expected, the fresher the food, the more it weighs. The research team hypothesized a division-level task force of five BCTs and their enablers, or 19,283 personnel, and then estimated their food requirements for a year (21.1 million meals). As shown in Table 3.1, if fed MREs, the force would require 12,220 tons of these rations (1,111 M977 HEMTT loads). A B-ration diet would weigh 17,596 tons (1,600 M977 loads). From these data, we calculate a 44-percent convoy premium for fresh food over MREs.

An A-ration diet for the force would weigh 19,283 tons (1,753 M977 loads) and a 58-percent convoy premium over MREs. Stated another way, if we assume one soldier casualty for every 20 convoys and 25 vehicles per convoy, this task force can expect one to one and a half additional casualties per year as a result of shipping B-rations or A-rations, respectively, over MREs.

Table 3.1
Comparison of Logistics Burden, by Meal Type

Meal Type	Annual Food Weight Required, in Tons	M977 HEMTT Loads
MREs	12.2	1,111
B-rations	17.5	1,600
A-rations	19.2	1.753

SOURCE: NSRDEC, 2012.

In reality, DFACs at large FOBs feed fresh food, often four meals per day. Packaged rations tend to be consumed during operations and while manning COPs where security and LOCs might be more contested. Consumption and spoilage drive replenishment rates. Members of 377th Theater Sustainment Command interviewed for an earlier project reported that all of the convoys they operated in Iraq involved delivery of rations. Of course, soldiers must always be fed, no matter where they are stationed, so food deliveries are unavoidable. The best mix of ration types is an area that could yield reduction in the logistics burden that QoL creates. However, our research found that access to fresh food is an important contributor to the soldier's perception of QoL. Further analysis needs to be done to optimize the mix of meals, but clearly soldier security demands a reassessment of this approach.

There is also some evidence that the presence of retail restaurants can have a significant effect on the demand for water. The research team calculated a "water penalty" for retail meals by estimating the additional water consumption associated with each retail restaurant meal prepared.[4] The water penalty estimated to accompany a retail restaurant meal prepared in Afghanistan is 1.5 gallons per meal (above that needed for A-rations prepared in a DFAC or in a force provider's field kitchen).

Figure 3.1 suggests that the potential savings could run to hundreds of tankers' worth of water per camp per year for theaters with water limitations similar to those found in Afghanistan and Iraq.

Given that many of the larger contingency bases in Afghanistan operate multiple retail restaurants, the additional water demand they generate could be substantial. Table 3.2 makes

Figure 3.1
Retail Meals and Additional Water Demand

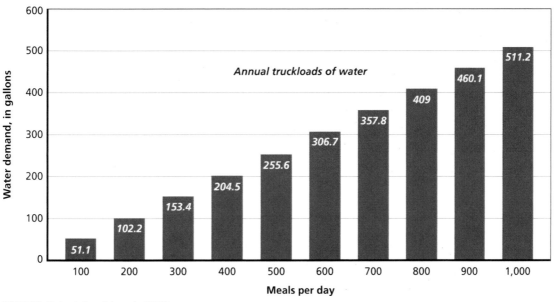

SOURCE: Dziegielewski et al., 2000.
RAND RR1298-3.1

---

[4] We used data and techniques from Dziegielewski et al., 2000. Dziegielewski and his colleagues placed meters on water sources in various restaurants around the United States and characterized food-preparation consumption from the meter data.

Table 3.2
Restaurants in the Base-Camp Sample

Camp Population	Camps in Sample	AAFES Restaurant Allocation	Total Restaurants	Water Consumption, in Trucks per Year[a]
1,000–2,000	9	1	9	1,381
2,000–3,000	1	2	2	307
3,000–4,000	2	2	4	614

NOTE: AAFES does not run every restaurant on a base camp. Examples of non–AAFES-run restaurants are Starbucks and KFC.

[a] Assuming 300 restaurant meals per day.

the matter more concrete. As the table illustrates, even modest levels of consumption at a small number of camps can produce hundreds of additional tanker-loads of water per year.

The sustainment community should work with the COCOM logistics staff to develop a policy on retail restaurants and similar concessions on contingency bases. The policy should be grounded in the supportability of their water requirements under the conditions prevailing in the theater of operations.

Water

Much of the demand for water is driven by population size and is based on the biological need for hydration and interest in showers, laundry, and toilets. The demand that requirements for human consumption generate might not yield substantial savings in water requirements. However, one practice, the use of bottled water, appears to increase overall distribution burden. Bottled water is a difficult commodity to handle. Palletized loads tend to come apart, rendering unloading via material-handling equipment unsuitable. There is a fair amount of loss and breakage associated with loading and unloading. Soldiers tend to discard bottles before they are empty and when the contents become warm and unpalatable. Heavy use of bottled water could inflate demand by 20 to 30 percent, according to some sources (Moore, 2011). However, the convenience of throwing a case of bottled water in the back of a high-mobility multipurpose wheeled vehicle and the ease of water availability do mitigate the risk of dehydration casualties, an important consideration in arid AORs. Balancing the logistics burden of bottled water with the ease and abundant access to water it provides might require a change in training to reinforce the use of hydration systems or, alternatively, investment in more-robust distribution methods for bottled water.

Another area of potential savings is in the water used for showers. An estimated 75 percent of the water required at contingency bases is utilized for showers. Since 2011, the Army has fielded 54 shower water–reuse systems (SWRSs) to units in Afghanistan:

> SWRS combines the tactical water purification system and hospital containerized batch laundry capabilities to treat and return to use up to 9,000 gallons of water a day. Using just one system at its full capacity can result in saving potentially 3.2 million gallons of water a year. ("Shower Water Reuse Systems Employed at Forward Operating Bases in Afghanistan," 2012)

There are other opportunities to create efficiencies and expand the options for providing water. Few theaters of operations are as arid as Afghanistan and Iraq. In many, it will be possible to draw water from rivers and lakes. Even in Afghanistan, wells are possible (e.g., at Kandahar Airfield), though they typically require U.S. drilling sets to reach the deeper aquifers, because U.S. practice recommends that the Army reserve the water nearer the surface for indigenous agricultural uses and human consumption (HQDA, 2015). In future operations in coastal areas, desalination of seawater might be possible if the LOCs are adequately secured. For example, Djibouti, on the Red Sea opposite Yemen, has such a plant. Depending on the threat and geography, pipelines could connect major consumers to sources, or at least substitute for surface distribution between major contingency bases. In climates like those of Afghanistan and Iraq, recovering, treating, and reusing wastewater is an alternative way to achieve potential savings. In addition to the SWRS mentioned previously, the Army is investigating these opportunities, including gray-water recovery from laundry and food preparation and similar sources. Other sources under investigation for reclamation include discharge vapors associated with engine, motor, and fuel cells. The National Aeronautics and Space Administration currently achieves 93-percent efficiency in recovering and reusing water aboard the international space station. Table 3.3 illustrates the research team's estimates of water savings based on the 2014 U.S. Army troop strength in Afghanistan (32,720 soldiers), assuming a requirement for 60 gallons per person per day and 30-, 40-, and 50-percent reuse.

Energy

Base-camp energy is an area rich in potential savings,[5] both in terms of electricity and in terms of fuel. We begin by considering fuel-saving options. One potentially very effective fuel-saving option is the selection of soldier housing and work location structures. Many options are available, from general-purpose medium tents to semi-permanent military construction (MILCON) at the other end of the spectrum. In between these two extremes are numerous prefabricated housing solutions. Each of these requires different amounts of fuel to achieve the same climate-control and electrical capabilities. In many cases, the differences are remarkable. One illustration of this effect involves replacing B-huts with structural insulated panel (SIP) huts. According to the research team's contractor-provided consumption reports, total monthly contingency base energy consumption for the 16 camps in our data set totals 1.4 million gallons of fuel. Moving to SIP huts could save as much as 504,000 gallons of fuel per year.[6]

Another fuel-based saving option is to move from spot generation to grid-based electricity. According to Vavrin et al., 2013, a power grid can reduce contingency base fuel use

Table 3.3
Estimates of Savings from Water Reuse

Percentage Recovered	Tanker-Equivalents Saved Monthly
30	6,690
40	8,910
50	11,160

---

[5] The analysis and estimates presented in this section draw heavily on data in Vavrin et al., 2013, and on contractor-provided reports of installation energy use for 16 base camps over 39 months.

[6] If we assume that 90 percent of that goes to living, dining, and office space provided in B-huts, the total would be approximately 1.2 million gallons. Given that SIP huts' energy demand is 60 percent of B-hut demand because of SIP huts'

between 10 and 50 percent. Cast in terms of the research team's 16-camp sample, on average, these savings could amount to 408,356 gallons per month of fuel.

Energy savings found in electricity generation include (1) exchanging the Vietnam-era tactical quiet generators for today's modern Advanced Medium Mobile Power Sources (AMMPS) systems and (2) moving from spot generation to grid-based power.

In a direct exchange of tactical quiet generators for AMMPS, such a move would affect 4,826 generators in Afghanistan producing 60 kilowatts (kW) or less. The improved fuel efficiency (an improvement of 0.254 gallons per hour) would save 882,579 gallons per month (334 tanker-equivalents) across the fleet of generators.

Relative Investment, by Phase of Operations

In this context, *relative investment* refers to the qualitative assessment of direct demand for the sustainment community to deliver goods and services, by phase of military operations. Generally speaking, the level of commitment from the sustainment community is a function of the mission requirements and therefore varies with the phase of operations U.S. forces are conducting. Table 3.4 summarizes QoL sustainment considerations.

Phase 0: Shape the Environment

A review of the Theater Security Cooperation Management Information System indicates that a majority of security-cooperation events involving U.S. Army troop units are short duration, expeditionary, and austere in nature. The participating units deploy with their organic QoL capabilities into a partner-provided location for bed-down. The United States maintains forward-stationed units for shaping and presence activities, and these instances (e.g., Saudi Arabia, Kuwait, Philippines) deserve special consideration for QoL matters. In these cases, the partner might prefer to isolate U.S. personnel from the local population (e.g., as Saudi Arabia did in

Table 3.4
Quality of Life, by Phase of Operations: Sustainment Considerations

Phase	Types of Operation	QoL Level	Design Consideration	QoL Saving
0	Security cooperation	Basic	Relations with partners	Contract support
I	Short term	Basic	Threat; force protection	Contract support
	Long term	Enhanced	Relations with host; morale and welfare of the force	Contract support
II and III	Short term	Austere	Sustain combat capabilities	None
	Long term	Basic	Sustain combat capabilities	None
IV	Short term	Basic	Expectation for transfer to civil authorities	Set policies for supportability by commodity consumed
	Long term	Enhanced	Supportability; perceptions of local population	Set policies for supportability by commodity consumed

superior insulation, the reduced energy demand would be approximately 504.000 gallons, or an estimated monthly savings of about 48,000 gallons.

1990 at remote military cities and bases). In other instances, such as Joint Special Operations Task Force—Philippines in the Philippines, force-protection considerations can limit freedom of movement during off-duty hours. In other instances, perhaps in future African partners, the local economy might be so poor that very little is available in it that adds to soldier QoL. In such cases, QoL could be critical to the morale and well-being of the force. The types and levels of QoL delivered should be calibrated to tour length: The longer the tour, the more abundant the QoL support. Personnel on yearlong tours at Camp Lemonnier in Djibouti have access to a swimming pool, USO shows, sporting and cultural events, and a wide variety of activities for the entire population, according to pictures on the camp's Facebook page.[7] Facilities in Kuwait are equally well developed.

Phase I: Deter

Because deterrence operations usually take place on friendly territory, contractor support is one way to calibrate QoL support. In the short term, force protection and threat considerations can impose limits. For example, if indirect fire and air attacks are a possibility, then DFACs and similar places for gathering might not be prudent, and tactical field-feeding postures might be preferable. As the confrontation matures and stabilizes over the long term, commanders might adopt a less defensive posture, and QoL facilities, such as DFACs, exchanges, theaters, and MWR facilities, might become feasible.

Phases Two and Three: Seize the Initiative and Dominate

QoL is generally minimized in Phases II and III because the mission and operating tempo (OPTEMPO) are often incompatible with enhanced QoL levels. If these phases grow longer, the demand for QoL will likely increase, and the sustainment community will have to be prepared to provide rest areas, bath and direct exchange (DX) points, and similar services.

Phase IV: Stabilize

Phase IV might have the greatest potential to reap savings in terms of QoL. The key to doing so involves developing policies that will constrain the otherwise nearly inevitable expansion of QoL through engineering improvements, MWR, and AAFES expansions and, with them, increased demands for facilities, fuel, energy, and water. The first order of business is to answer this question: What level of effort is sustainable in each commodity area (e.g., numbers of tankers per month for fuel and water), given the other demands on the sustainment community and the characteristics of the theater?

In some theaters, LOCs will be more abundant, and multiple, independent supply convoys can be used to resupply all the outposts being supported. Where capacity is still short of requirements, it might be possible to re-task convoys that finish their routes sooner (probably the shorter routes) to assist in delivering supplies to the remaining outposts that have not yet been resupplied. In instances in which this approach is effective, the Army can generate extra efficiency in its available capacity and can avoid rationing. In other cases, in which capacity still falls short, commanders can impose rationing. Food is one of the easier commodities to manage. As Chapter Two pointed out, by shifting from A-rations to a mix of MREs and B-rations, the Army can conserve cargo space and payload because the fresher the rations, the

---

[7] See Camp Lemonnier Djibouti's Facebook photo stream.

more they weigh and the more space they consume. Thus, moving away from fresh rations can ease the load on the available delivery capacity.

In terms of water, the theater can shorten shower time at the margins (e.g., 30 to 60 seconds) per person. In non-arid theaters, it might be able to substitute non-potable water for some flight-line and motor-pool operations, depending on local circumstances. In terms of contingency base energy fuel, the theater can ban the use of personal electronics and ration MWR and AAFES electricity consumption.

Another key consideration is standardization of base-camp equipment, especially generators. All other considerations being equal, standardization facilitates accountability (e.g., fewer line items to oversee), simplifies maintenance and repairs (e.g., fewer spare parts to maintain, fewer specialty tools required to do the work), and requires fewer technicians to maintain the infrastructure.

If it becomes clear that Phase IV will be short lived and that control will pass to civil authorities soon, the sustainment community will be increasingly engaged in the recovery and movement of Army assets to ports of embarkation for return to home station. Its capacity for QoL support might, therefore, be more limited. If, however, residual resistance persists and U.S. engagement is expected to continue, QoL policy might be relaxed somewhat—although still bounded by supportability constraints—to maintain morale and resiliency within the force.

Savings Associated with Soldier Preferences

If future soldier preferences for QoL goods and services prove consistent with the preferences reported in the project's crowdsourcing effort,[8] there could be other, significant opportunities to reduce QoL requirements. The top 80 percent of QoL demand consisted of hot meals, laundry service, running water, good communications home (including internet access), a gym, privacy, solitude, and restful sleeping conditions (i.e., climate, noise, and light controlled). MWR services, showers, AAFES and PX, and "good DFAC" completed the list.

Fairly rudimentary base-camp features can satisfy these demands. The internet connectivity presumes electricity and appropriate providers, but these conditions have been satisfied in the recent past.[9] Running water might require access to a well or reservoir, or perhaps a pipeline. Privacy, solitude, and conditions conducive to sleep can be achieved by building B-huts sufficient to give the entire force one-person rooms. For the force deployed in Afghanistan during 2014 (32,720), the bill would be approximately $61.4 million. Using SIP huts might cost marginally more, but energy savings (and thus fuel savings) from the SIP hut's superior insulation would offset the cost difference. Climate control also improves with SIP huts.

By satisfying the top 80 percent of QoL demand, the sustainment community would also address some aspects in each of the main five categories in Maslow's hierarchy of needs. These include physical and material well-being, relations with other people, recreation, social and community activities, and personal growth and fulfillment (Kerce, 1992).

---

8   See Appendix B for complete details.

9   At one point, Camp Leatherneck had installation-wide Wi-Fi and supported more than 6,000 users.

CHAPTER FOUR

Recommendations to Improve the Efficiency of Quality-of-Life Support

This chapter recommends efforts the Army should consider toward the goal of future expeditionary contingency bases. These efforts contribute to one or more of the benefits to effective base-camp management discussed earlier in this report: (1) enhanced personnel security, (2) improved energy security, and (3) better cost efficiency. We close the chapter with an assessment of the detailed near-term priorities required to better inform our recommendations.

Recommendations

We have chosen to organize the recommendations by DOTMLPF. This serves to group the recommendations generally by the Army Staff, secretariat, ASCC, or direct reporting unit that we determine has the lead responsibility in addressing the recommendations.

Doctrine

- Maximize the use of local water sources in support of base-camp operations. Joint doctrine mandates that tactical bulk-water support operations should purify water as close to the user as possible. The Army Corps of Engineers maintains a geospatial database of water sources worldwide. This includes not only lakes and rivers but also groundwater sources and estimated volumes of water. These local sources reduce the need for water convoys and will increase sustainment security as a result. Planners should assess local potable water sources as a selection criterion when conducting base-selection COA analysis.

- Give particular emphasis to integration of water-recovery technology into base-camp operations and to a reassessment of water quality that maximizes the uses for recovered gray water. Base-camp water management does not end after its initial consumption, with a significant volume of gray water and black water generated. Typically, wastewater has been simply removed from the base camp via contracted assets, resulting in additional expense and attacks to the gray water–removal convoys. The Army has fielded the SWRS, capable of efficient expeditionary gray-water processing, and continues to develop expeditionary water-recycling and reclamation capabilities. Doctrine should continue to evolve to institutionalize an expeditionary focus and the emerging equipment capabilities.

- Update existing base-camp doctrine to institutionalize the lessons learned since the most current version was published in 2013. Areas that should be addressed include more-detailed QoL management processes, a focus on the "expeditionary" base camp described in multiple Army long-range strategic documents, and the coordination of effort required to deliver QoL at a sustainable level consistent with the theater commander's operational priorities.

- Supplement existing documentation with doctrinal information and standards, forming a single base-camp development regulatory standard. Interviews with ASCCs indicate that COCOM regulations on construction and base-camp development are integral source documents in the initial base-camp planning process. These regulations, also known as the Sand Book in CENTCOM and known by other colors in other COCOMs, are a rich source of information for the base-camp planner. However, these documents focus on roles and responsibilities, with some discussion of development processes. Nothing in the Sand Book or its sister regulations provides any planning factors or any other means to calculate an order-of-magnitude sustainment-demand estimate. This is a significant shortcoming. Many of these planning factors do exist, but they are found in multiple doctrinal products. The effort to research these planning factors and include them in base-camp development regulations is worthwhile because this should result in better base-camp life-cycle decisions.

Organization

- Consider creating MTOE organizations to construct and administer contingency bases. Doctrinal responsibility for the various aspects of Army base-camp construction, development, and management is distributed across multiple organizations, requiring a coordination of effort that yields mixed results. Other services, particularly the Air Force, have achieved success in standardizing development through the use of dedicated MTOE organizations to deliver contingency bases.
- Assign base-camp management responsibility to an MTOE organization for contingency bases of fewer than 6,000 residents. Regional support groups are doctrinally assigned the commander responsibility for contingency bases with populations above 6,000. However, the vast majority of contingency bases are below that population level. For these smaller contingency bases, units are assigned commander responsibility haphazardly, generally with little or no training on managing base-camp operations. Designating a unit the doctrinal responsibility for smaller bases will result in better-trained commanders, resulting in contingency bases more effectively and cost-consciously managed.

Training

- Institutionalize base-camp management training. The great majority of base-camp commanders come to that responsibility with little or no specific training for the role. The Army Office of the Assistant Chief of Staff for Installation Management (OACSIM) conducted the only base-camp management training of which we are aware. OACSIM developed a base-camp management curriculum for a 40-hour block of instruction in 2009, based on lessons learned from Iraq and Afghanistan. Although the OACSIM effort was admirable, its effectiveness was limited because it was developed outside the Army's training and doctrine organization. The lesson plans that OACSIM developed should be used as a starting point for TRADOC to develop a base-camp management training program that can be maintained in accordance with existing regulations and practice.

- Develop and institutionalize training and tools that enable COCOM and ASCC planners to more effectively conduct base-camp life-cycle planning. Training would prepare planners to make key decisions about required sustainment levels, based on the availability of food, water, and fuel in the theater and the transaction and opportunity costs that accompany them. Decision-support tools that help planners develop base-camp COAs should supplement this training. These tools should include cost–benefit analysis modeling capability and a standards-based menu of options that set a baseline for sustainment requirements.

Material

- Expand the Army's AMMPS inventory as a replacement for other tactical generators currently in use. AMMPS generators can be linked together to form tactical microgrids, which are demonstrably more efficient than single generators.
- Continue to build capability for base-camp water management. Programs of record exist for current and future solutions for base-camp management of wastewater recovery, treatment, and reuse, as well as potable-water storage and packaging compatible for smaller contingency bases. These solutions all contribute to sustainment security by making contingency bases more self-sufficient for water and therefore minimizing the convoys required to bring bulk water to the contingency bases and remove water.
- Develop and implement minimum energy-efficiency standards for temporary structures. This report has discussed at length the impact of shelter energy efficiency on base-camp energy and sustainment security. The Army should establish a minimum energy-efficiency standard for procurement and development of temporary structures that ensures that the benefits of energy efficiency can be sustained into future acquisition cycles.

Policy

- Develop policies that constrain automatic expansion of base-camp facilities and services in support of the combatant commander's operational priorities. Recent history demonstrates that base-camp commanders and tenants arrive at a base camp at the beginning of their rotations and strive to "improve their fighting position" in the name of "taking care of soldiers," as the Army commonly says. This improvement, although well intentioned, often serves to exacerbate the sustainment burden with limited incremental improvement in QoL. Senior leaders should communicate clear guidance to subordinate commanders regarding the acceptable limits for base-camp improvements.
- In tandem with the foregoing, establish, and promulgate widely, standard levels of support for base-camp services and sustainment for each phase of the base-camp life cycle. These levels of support provide a menu-of-options starting point for COCOM planners and base-camp commanders to appropriately sustain a base camp, based on the base-camp life cycle. This provides a baseline by which the base-camp commander can prioritize delivery of sustainment goods and services. The menu of options should be communicated throughout the Army at every opportunity, thereby establishing the new standard for base-camp QoL and managing soldier expectations.
- Establish energy-efficiency requirements or incentives in base-camp energy support contracts. A significant number of contingency bases receive power generation through contracted support. Few of these provide any efficiency or consumption standards that incentivize the contractor to limit its fuel consumption. These contracts should also stipulate

that fuel-consumption and energy-efficiency data that the contractor gathers be made available to the Army, unencumbered. Data gaps resulting from proprietary collection of energy data by contractors not being turned over to us have hampered the research for this report. This same information is essential to the Army for purposes of assessing energy efficiency at individual contingency bases, as well as assessing best practices across a network of contingency bases.

- Eliminate or limit retail restaurants and thereby avoid the water penalty associated with meal preparation and cleanup. As the estimates in Chapter Three indicate, the water penalty can quickly generate additional tankers' worth of water requirements.

Facilities

- When planning, take into account the limitations of minor MILCON thresholds with respect to energy-efficient semi-permanent and permanent base-camp construction. At present, legal thresholds limit the use of operations and maintenance dollars normally to $1 million per project.[1] Beyond this threshold, MILCON dollars are required, which must be programmed and are not quickly available during contingencies. The minor MILCON threshold can be insufficient for building generator microgrids to heat and cool facilities in the theater because, if the generators are networked, the sum of the individual units counts toward a single project cost, depending on the legal interpretation of what constitutes a project.
- Develop and implement minimum energy-efficiency standards for semi-permanent and permanent base-camp structures.

Looking Forward: Quality of Life in the Future Expeditionary Army

If the Army wants to pursue a leaner, more expeditionary approach to contingency operations than that followed in the recent past, it might begin by conceiving of multiple QoL packages, each optimized for very different circumstances.

Quality-of-Life Packages

As the analysis in the previous sections of this report makes clear, food, water, and fuel underpin virtually all QoL goods and services and collectively represent most of the distribution demand on the sustainment community. Fuel provides the contingency base with energy for everything that runs on electricity. Water is essential for life, food preparation and cleanup, laundry, sanitation, and medical support. Food is essential for morale, health, and fitness. Given the centrality of food, water, and fuel to operational success and to QoL, supportability becomes an important consideration. The transaction costs and casualties associated with delivering QoL goods and services (e.g., interim storage and transportation costs) must be minimized to the greatest extent possible, given the circumstances in the theater. Opportunity costs (i.e., activities that the sustainment community defers in order to deliver QoL-related goods) must also be kept to acceptable levels. Supportability in this sense should trump QoL standards, although standards remain important for establishing expectations for QoL delivery.

---

[1] Some exceptions apply that require secretarial approval and congressional notification or pertain to immediate safety requirements (10 U.S.C. § 2805).

Contingency bases are the primary mechanism for delivery of QoL goods and services. The specifics of delivery vary depending on whether the environment is non-permissive or permissive.

Non-Permissive Theaters

Army tactical units would become responsible for much of the QoL delivery in non-permissive theaters. This approach is consistent with the G-4 vision for lean expeditionary forces, able to fend for themselves with minimal outside support. It departs from current practices, however, requiring that MTOEs expand to provide company-level units with appropriate tents or shelters, food preparation, and field sanitation capability to operate in a non-permissive area of operations. These assets must be organic to the units so that soldiers can train with them and become expert in their operation and maintenance. For instance, combat service support (CSS) units would have to be capable of providing these services so that units on the move might pass through them in the normal course of operations (e.g., a unit is given relief in place and uses the break from operations to visit a bath and DX point to clean up). The usual practice of establishing in-theater R&R sites would supplement unit efforts to address QoL needs.

Base-camp equipment sets could also be issued to the corps-aligned expeditionary sustainment commands and to the division- and corps-aligned combat sustainment support battalions.[2] Thus equipped, these units could begin erecting contingency bases as soon as the theater commander's requirements for their number, locations, and capabilities begin to emerge. The equipment sets could be configured for a battalion task force (e.g., approximately 1,000 troops) and a BCT with enablers. Another package might be designed to support a major headquarters (e.g., division).

Permissive Theaters

As soon as conditions are safe enough for contractors to operate, the QoL portfolio goes to the contractors establishing and operating the campaign's fixed sites. Many of the key sites will have been established at this point, and the tasks for contractors will be building out the existing camps; replacing unit MTOE tentage with sturdier, more energy-efficient temporary buildings; and establishing facilities for a DFAC, MWR, a gym, and other QoL amenities consistent with QoL policy. In some instances, requirements for entirely new sites will be identified, including sites where large concentrations of U.S. forces and multitenant contingency bases are desirable. In this process, commanders should interpret QoL policy for each camp, old and new alike.

Supportability remains a key determinant of QoL policy implementation. The key to controlling QoL demand is to set sound policies to monitor and control the consumption of food, water, and fuel.

---

[2] For a description of these sustainment units, see Mason, 2013.

Base-Camp Case Studies from Southern Afghanistan

This appendix describes an analysis of the QoL characteristics at contingency bases in Afghanistan from June 2012 to March 2014. The hypothesis that the relative lack of standardization in QoL characteristics among contingency bases and FOBs across the theater created challenges for the logistics distribution system to meet demands motivated this case-study analysis. Because QoL ranks high among commanders' priorities, nonstandard requirements can create inefficiencies that can lead to requirements for larger bases, increased numbers of contractors on base and outside the wire to provide the services, and greater overall costs to sustain these base-camp configurations under shifting populations and operational circumstances. Much of the information discussed here is qualitative and descriptive. However, wherever possible, we try to quantify those criteria or characteristics being analyzed.

Methodology

This analysis relied primarily on two sets of data: reports required of contractors participating in LOGCAP and a U.S. Army Logistics Innovation Agency (LIA) Base Camp QoL Enhancement Cost Estimator model. The QoL Cost Estimator is a spreadsheet-based tool developed at the request of G-4 to provide logisticians, camp commanders, and other interested parties with a way to estimate the cost of QoL on contingency bases based on published standards for QoL requirements. Several other data sources augmented these two sets of data as needed.

Two LOGCAP task orders covered QoL support from contractors in Afghanistan: the Fluor Corporation task order in the north and the DynCorp International task order in the south. We focused on contingency bases located in southern Afghanistan for two reasons: Most of the U.S. Army active forces were located there, and operating conditions changed routinely, which complicated the Army's ability to sustain QoL standards. The type of LOGCAP data analyzed pertained to base-supported populations (BSPs), types of facilities, and power-generation capabilities.

Background

At the beginning of OEF, the United States collaborated with Afghan forces to take control of the main cities in the north and eventually the capital, Kabul. Once the main cities were secured, U.S. forces operated primarily out of Bagram near Kabul and Kandahar further to the south. FOBs and COPs were constructed from green fields throughout the country to avoid disturbing local populations in their cities and villages and their respective farming activi-ties. North Atlantic Treaty Organization (NATO) and coalition forces operated primarily in northern Afghanistan in the mountainous regions. U.S. forces operated primarily in southern Afghanistan near Pakistan and in western Afghanistan. A mixture of all three could be found throughout the theater.

Figure A.1 shows the southern region of Afghanistan. The square outlines those bases included in the southern Afghanistan task order. The areas in white ovals are the four bases on which we focus in particular: Kandahar, Camp Leatherneck, Spin Boldak, and FOB Lindsey. Kandahar and Camp Leatherneck are large bases. Spin Boldak is medium sized, and Lindsey is a small base. We selected them on the basis of their size and whether they were increasing or decreasing in population over time. LOGCAP data reporting is very detailed, and analyzing bases over time required that we focus on a select few rather than all.

Most of the contingency bases in Afghanistan are relatively small. Figure A.2 shows population size ranges and numbers of bases within these ranges. The ranges represent the average population size of the bases over the reporting period of June 2012 to March 2014. According to the LOGCAP program manager at Army Contracting Command (ACC)–Rock

Figure A.1
Case-Study Contingency Bases in Southern Afghanistan

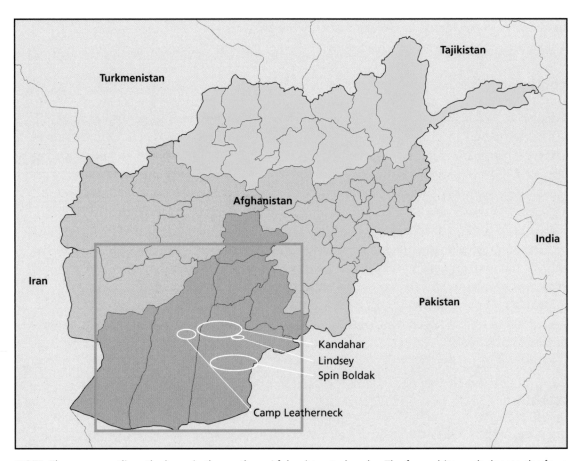

NOTE: The square outlines the bases in the southern Afghanistan task order. The four white ovals denote the four bases on which we focus in particular: Kandahar, Camp Leatherneck, Spin Boldak, and Lindsey.
RAND *RR1298-A.1*

Figure A.2
Contingency Base Sizes in Afghanistan

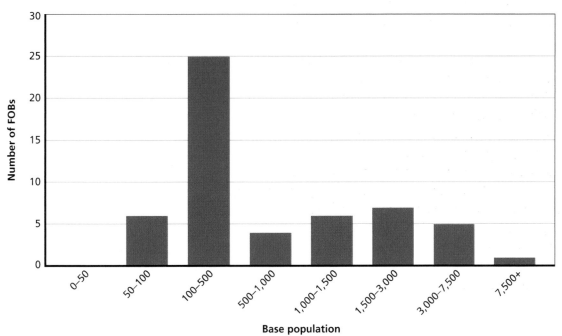

SOURCE: ACC–Rock Island data.

RAND *RR1298-A.2*

Island, population counts are difficult to determine. Maneuver units moved in and out of bases frequently as they deployed in and redeployed out or as operations required. Units in outlying, austere areas that had rudimentary QoL standards would also come into larger bases in the rear to take advantage of higher QoL benefits, such as showers, laundry, and semiprivate billeting quarters.

An accurate count of the number of personnel on a base camp is important for logistical planning and because it is the means by which LOGCAP contractors are compensated for services rendered. The best way of estimating the number of personnel on a base at any one time is to count the number using the DFACs. Personnel are tagged to ensure that they have DFAC privileges, for example. Local contractors are not permitted access to the DFAC, but LOGCAP contractors and others living inside the wire are.[1] However, even this approach had its flaws in that it could not distinguish between personnel visiting and eating at the DFAC from those living on the base camp.

According to Figure A.2, most of the bases in southern Afghanistan had between 100 and 500 personnel. A handful of bases were still smaller—with populations of 50 to 100—or larger, with populations of more than 3,000.

Figure A.3 shows the four bases discussed in more detail in this section and their respective BSPs as reported in LOGCAP and ACC–Rock Island data. The decision to surge U.S. forces was made in December 2009. Beginning in 2010, more soldiers, airmen, and marines deployed to south Afghanistan. More recently, the net number of U.S. forces has been draw-

---

[1] *Inside the wire* is a colloquial Army term describing the area inside a perimeter defense. That can be a base camp for semipermanent structures or simply a defensive perimeter for a COP.

Figure A.3
Base-Supported Populations for the Four Case-Study Afghanistan Bases

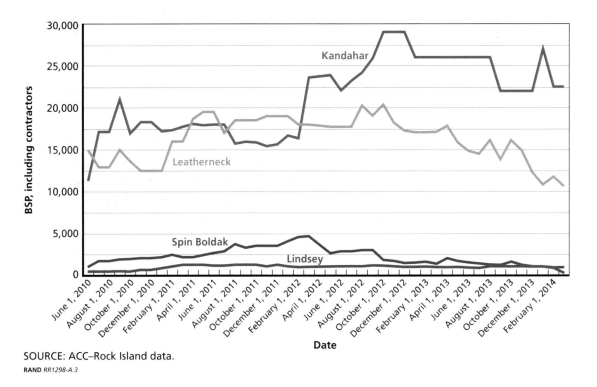

SOURCE: ACC–Rock Island data.
RAND RR1298-A.3

ing down as they have redeployed out of the theatre. Kandahar and Camp Leatherneck have populations of more than 10,000 and 25,000, respectively, the largest in the area. Spin Boldak had the largest fluctuation in its population, reaching as many as 4,700 in mid-2012 before receding to approximately 1,000 by early 2014. Lindsey was the smallest, at slightly more than 1,000 or below. However, because Lindsey was located close to Kandahar and had a DFAC, some units rotating in and out of theater visited Lindsey, making it more difficult to estimate the number of permanent personnel supported. These transiting personnel could have boosted its numbers to more than just the resident population.

Quality-of-Life Standards at Four Contingency Bases

The project team classified facilities associated with QoL at four contingency bases in South Afghanistan: Kandahar Airfield, Camp Leatherneck, Spin Boldak, and FOB Lindsey. The number of data files required scoping the analysis to a subset of bases that spanned a range of populations. Kandahar and Camp Leatherneck were the largest bases in southern Afghanistan and offered the largest selection of QoL types and levels. Spin Boldak was a medium-sized base, and Lindsey was a small base.

On the four contingency bases in the analysis, the majority of real property and buildings (an average of 80 to 90 percent) are identifiable as QoL-related facilities (mainly housing). A small percentage of items are identifiable as non-QoL facilities. Non–QoL-related buildings pertained to military, operational, administrative, and contractor activities, such as guard towers, supply and maintenance warehouses, maintenance facilities, Defense Contract Management Agency facilities, Defense Reutilization and Marketing Office facilities, and office buildings. We could not identify approximately 10 to 20 percent of facilities. Figure A.4 shows these proportions in number of buildings (rather than total square footage) for June 2012, a month chosen for the descriptive analysis based on its low proportion of unidentifiable facilities or higher proportion of filled-in data.[2] In Figure A.4, "Other QoL related" includes categories that are related to life-support activities other than billeting. This includes AAFES facilities; personal services, such as automated teller machines; DFACs; MWR buildings; gyms; LSS buildings; and laundry facilities. An "unidentifiable" building is one that had missing data pertaining to its purpose or function.

The project team identified QoL-related property and characterized each facility by its primary purpose and by its primary type of building construction material (e.g., tent, hard stand). We based the QoL categories, facility types, and structure types on Army engineering standards, as represented in LIA's QoL Cost Estimator. Table A.1 shows the major QoL categories and subcategories, and Table A.2 shows the building structure types.

Of the facilities identified as QoL related, the overwhelming majority (approximately 70 to 90 percent) are housing; 7 to 24 percent are latrine, shower, or laundry; 1 to 3 percent are recreation; less than 1 percent personal services; and less than 1 percent food service. Figure A.5 shows these proportions, in number of buildings, for June 2012. Again, we chose this month based on its low proportion of unidentifiable facilities.

Figure A.4
Proportion of Buildings, by Quality-of-Life Support Type

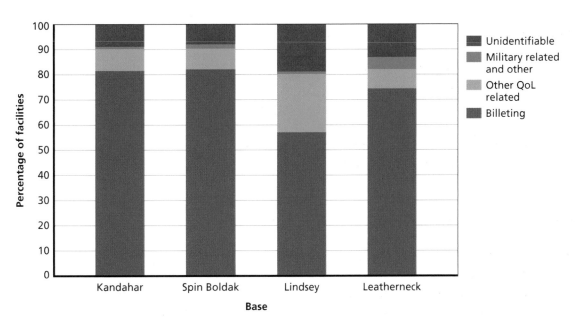

SOURCE: ACC–Rock Island data.
NOTE: "Other QoL related" includes categories that are related to life-support activities other than billeting. This includes AAFES facilities; personal services, such as automated teller machines; DFACs; MWR buildings; gyms; LSS buildings; and laundry facilities. An "unidentifiable" building is one that had missing data pertaining to its purpose or function.
RAND RR1298-A.4

---

[2] In other months, unidentifiable facilities can make up a much larger proportion of the installation real property—in some cases, more than 87 percent.

Table A.1
U.S. Army Logistics Innovation Agency Cost Estimator Quality-of-Life Major Categories and Subcategories and Master-Schedule-of-Work Building Descriptors

LIA QoL Category	LIA QoL Facility Type	MSOW Descriptor
Food service	Coffee shop	Coffee shop; Green Beans (a particular coffee shop)
	DFAC	DFAC; dining facility; dining hall; dining tent
	Kitchen	Kitchen
Housing	Housing	Barracks; billeting; billets; housing; living quarters
Latrine shower laundry	Latrine	Latrine
	Latrine shower	Combo unit; shower/latrine
	Shower	Shower
	LSS	Hygiene unit; LSS
	Full-service laundry	Full laundry
	Self-service laundry	Self laundry
	Laundry pickup/drop-off	Drop off/pickup; drop off point; pickup point; drop off tent
	Laundry	Laundry (if no other identifiers)
Personal services	AAFES	AAFES
	Alterations	Alteration
	Barber	Barber/coffee shop (if no other identifiers); barber shop
	Beauty salon	Beauty shop; salon
	Bus depot	Bus; bus depot; shuttle bus
	Chapel	Ablution; chapel; chaplain; Christian ministry center; church; religious personnel; spiritual resiliency center
	Gift shop	Gift shop
	Post office	Mail; postal; post office
	PX	Exchange; post exchange; PX
Recreation	Call center	AT&T; call center; phones; phone room; telephone
	Education center	Classroom; education annex; education center
	Gym	Fitness center; gym
	Internet café	Computer room; internet
	Local vendor	Boardwalk tent; carpet shop
	MWR	Break room; conference room; cultural center; gaming; media center; morale center; MWR (if no other identifiers); rest facility; TV room
	USO	USO

Table A.2
U.S. Army Logistics Innovation Agency Cost Estimator Building Structure Types

Structure Type	MSOW Descriptor
B-hut/SWA-hut (wood)	Barracks hut (B-hut); B-hut/Southwest Asia (SWA)-hut; SWA hut
Brick	Brick
Concrete	Concrete armor unit; concrete masonry unit; concrete
Container	Containerized housing unit (CHU); containerized living unit (CLU); connex; container; modular; relocatable building
Generator	Generator
Hard structure	Hardstand; hard structure
Light set	Light set
Metal	Metal; CORIMEC (a company that makes prefabricated structures); K-Span (a machine that builds prefabricated buildings); pre-engineered building
Stone	Stone
Tent	Alaskan tent; big top; clam shell; hangar; TEMPER tent; tent
Wood	Wood

NOTE: CHU = containerized housing unit. TEMPER = tent, extendable, modular, personnel.

Figure A.5
Proportion of Facilities, by Quality-of-Life Category for June 2012

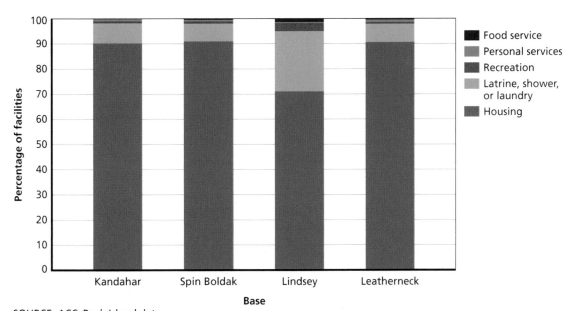

SOURCE: ACC–Rock Island data.

RAND RR1298-A.5

How Quality of Life Affects Logistics Requirements

The two classes of supply that place the greatest demands on the logistics distribution system are classes I (food and water) and III (bulk fuels and packaged petroleum, oils, and lubricants). They constitute recurring daily requirements, are heavy by unit volume, and have high volume requirements. Food and water requirements to be shipped to a base are a function of the BSPs and available well water. Fuel and petroleum requirements are a function of the contingency base power-generation needs and tactical operations requirements. Power-generation demands are a function of the number and type of generators used and facility types. Well-sealed buildings with thick walls, such as concrete facilities with windows, will use energy more efficiently than basic tents equipped with heating, ventilation, and air conditioning. Generators, not local utilities' power lines, supplied virtually all power that contingency bases in Afghanistan required. Contingency bases provide their own power to ensure consistent equipment and lighting operations and to avoid competing with power-generation needs of local populations. Local power generation in Afghanistan was relatively constrained and routinely experienced brown- and blackouts, also making it an infeasible source.

Class I: Food and Water

The military fellows we interviewed and the survey respondents considered the quality and quantity of food one of the most important QoL standards. Soldiers and airmen said that, if they were in austere locations or at the ends of their missions, they looked forward to having hot meals of familiar foods at the ends of their duty cycles.

Because class I supplies are common commodities required across all the services, the Defense Logistics Agency (DLA) is responsible for providing them. Class I requirements begin with menus that personnel at Fort Lee in Virginia determine. They develop menus for seven- and 14-day cycles. Commanders can choose menus for each meal and day for their respective bases. At the end of a cycle, these menus repeat until selections or cycles are changed. These choices are entered into DLA's data system, which is accessible to the direct-vendor delivery contractor. A single food supply contract was awarded to Supreme Foodservice in Switzerland from 2001 to 2010 and then to ANHAM in the United Arab Emirates from 2010 to the present.

Dry goods and refrigerated foods can be shipped in surface ships. Fresh vegetables and fruits are usually flown in by air because of their respective short shelf lives. Food vendors are responsible for providing prescribed days of supply of MREs and food supplies to fulfill menu requirements across the theater's bases. They are required to use the Defense Transportation System to transport food via surface ships or aircraft over long distances, such as from the continental United States or Europe. Vendors palletize their foodstuffs and deliver them to a DLA distribution center or port. From there, DLA ships the goods to the theater using its transportation account codes and later bills the Army for second-destination transportation charges. To reduce contract costs, the services, rather than the vendors, pay shipment costs directly.

Once food pallets have been delivered to class I warehouses at Bagram and Kandahar, the vendor picks them up and transfers them to its own warehouses. Food vendors are responsible for moving the food from their warehouses to the various contingency bases, typically subcontracting with local trucking firms to transport these supplies. ANHAM's website says that it operates a dedicated fleet of 250 heavy-duty trucks. Once at the base, trucks wait in a yard for one to three days to await safety inspections. Seals that the vendor installed at its warehouse are inspected to ensure that they have not been tampered with. When the trucks have been cleared, they enter the camps and deliver the food to the LOGCAP contractor or, if LOGCAP does not operate at the base, to camp personnel. At bases where LOGCAP contractors do operate, they are responsible for storing and distributing food and water once these supplies arrive inside camp premises.

Three water-production plants were built soon after U.S. forces were deployed in large numbers to the theater. Water is heavy and expensive to transport relative to its production value and is more economical to produce in the theater itself. These water-production plants are at Bagram, Kandahar, and Leatherneck and undergo routine health and safety inspections. Water produced here is used for bulk cooking and drinking and bottled-water purposes. Water wells within the wire at various camps provide supplies for laundry and latrine purposes.

Trucks in convoys transport most food and water. Some austere camps that did not have secure surface access or camps that ran into shortages received supplies by air via helicopters. Under the Supreme contract, the contractor provided air shipments. At the time of this research, air shipments were not conducted under the ANHAM contract. When an air shipment is required but conditions are too insecure for civilian traffic and under emergency conditions, military assets are used. However, these occurrences are relatively infrequent.

AAFES provides dining cafes and food items and is responsible for transporting its own food supplies into theater. It uses the Defense Transportation System and has its own transportation account codes. As a nonprofit activity, AAFES recovers its expenses by charging customers for the goods and services supplied. At larger bases, AAFES usually has facilities.

Concession operations, such as Burger King, TGI Friday's, and Baskin-Robbins, are responsible for bringing in their own food supplies. These companies can subcontract transportation of supplies to bases with the larger food vendors, though generally they operate only at the larger bases, where distribution distances are relatively short.

Class III: Fuel and Packaged Petroleum, Oils, and Lubricants

Class III supplies are another bulk commodity that DLA procures for all military services. Class III is used for both military operational requirements and for meeting QoL demands. Beginning in 2011, the Army Petroleum Center began an initiative to measure fuel consumption at up to 39 contingency bases. The initiative concluded in September 2014. Class III requirements are communicated via classified networks because of their relationship to OPTEMPOs and tactical operations. No central data system exists for administering fuel transactions. Instead, shipments, receipts, and vouchers are all managed manually. Accountability and management are challenging. Fuel shrinkage (downloaded supply volumes being less than what was uploaded) has been a continual issue. Several interviewees said that fuel was considered the second-best commodity for trade after money. All fuel must be imported into Afghanistan. It is in short supply in the domestic market and can be used by all sorts of vehicles.

Since 2003, CENTCOM has provided from 300,000 gallons of petroleum per day to more than 5 million gallons per day (Carra and Ray, 2010). As a land-locked country with no national oil-refining capability, Afghanistan must have all fuel transported into the country, distributed, and stored. According to Carra and Ray, including Iraq and other CENTCOM countries, contracts for more than 2,000 commercial trucks distribute fuel and manage 200 million gallons of contracted petroleum storage. To minimize transportation costs, the Defense Energy Support Center (DESC) of the DLA buys fuel primarily from commercial suppliers and refineries in the Gulf and Central Asian states, Turkey, and Pakistan (Carra and Ray, 2010). DLA competitively bids fuel contracts and requires contractors to transport the fuel into Afghanistan to a central depot supply point. The ground LOCs for petroleum delivery into the country are shown in Figure A.6. These fuel contractors are responsible for subcontracting to vendors to transport fuel via trucks, barge, and rail from neighboring countries into Afghanistan. Once the fuel has been delivered to defense central supply points in Afghanistan, CENTCOM and the Army assume responsibility for providing inland petroleum distribution to all customers (Title 10) (interview with Army Petroleum Center staff). These customers include U.S. and coalition forces; the Afghan National Army, Afghan Air Force, and Afghan National Police; U.S. civilian agencies; and contractors.

Figure A.6 shows fuel ground LOCs for Afghanistan. Figure A.7 shows the refineries that provided most of the fuel. In 2006, CENTCOM was importing about 16,000 gallons of petroleum per day. Turkmenistan provided about 2,700 gallons per day, Uzbekistan about 2,400 gallons per day, and Pakistan supplied about 10,800 gallons per day. DLA contracted with refineries in Pakistan, which supplied jet propulsion fuel, type 8; motor gasoline, an

Figure A.6
Central Asia Ground Lines of Communication for Fuel to Afghanistan

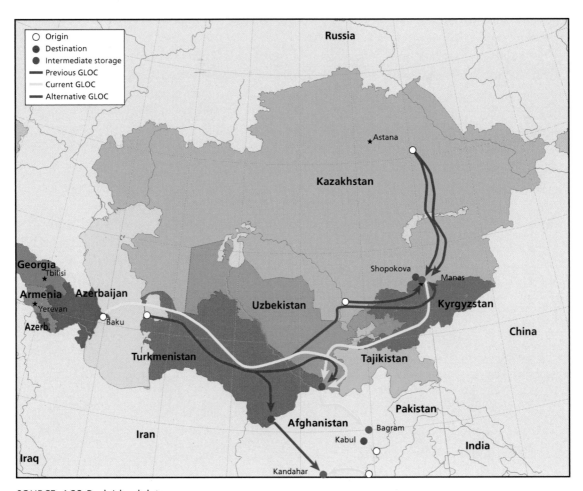

SOURCE: ACC–Rock Island data.

RAND *RR1298-A.6*

Figure A.7
Supply of U.S. Central Command Petroleum to Afghanistan from Refineries in Neighboring Countries

SOURCE: ACC–Rock Island data.
NOTE: kb = thousands of barrels.
RAND *RR1298-A.7*

unleaded fuel; and winter- and summer-grade diesel fuel; and with a private company that contracted with refineries in Turkmenistan and Uzbekistan for jet-grade fuel (Faust, 2007).

The Joint Petroleum Office establishes fuel requirements, which are communicated to DESC. DESC purchases the fuel on a competitive basis. Contractors have about a 21- to 30-day order and ship time to move petroleum from Central Asian sources to Bagram or Kandahar (Carra and Ray, 2010). Fuel comes into Afghanistan via rail just inside the border. From there, commercial trucks deliver fuel to their destination or storage points. From there, it is shipped to about 70 FOBs (Faust, 2007).

In 2006, most of the fuel came from refineries in Pakistan (Karbuz, 2006, citing a mission briefing given at the 2006 World Wind Energy Conference). However, in 2007, most of the fuel deliveries into country transitioned from the south to the north along the Northern Distribution Network after Pakistan shut down the southern border crossing.

Because Afghanistan operations include NATO, CENTCOM has leveraged assets and contracts in force from NATO's Allied Joint Force Command Brunssum to provide fuel support to U.S. forces in Regional Commands South and West. NATO contract costs are nonetheless payable by the Army, although these agreements are not recorded in formal Army data systems. During OEF, fuel transactions were manually processed, making it difficult to manage class III supplies and funds efficiently. Several interviewees posited the high likelihood that, when the Army received a fuel bill, having no independent recourse to seek reimbursement from other parties, it probably paid the bill regardless of the type of customer.

Having most of the fuel concentrated in two locations meant that distribution required all trucks to go through the main gate where all classes of supply pass. Additional gates and entry points became MILCON projects; truck prioritization was used in the interim period. Another bottleneck occurred within the gates when trucks waited in a yard until the download or upload equipment was available. Expanding this capability became another MILCON project (Faust, 2007). To relieve this traffic, some fuel was pushed out to locations prior to their demands.

Comparison of Quality-of-Life Standards: Contingency Base Facilities in Afghanistan and the U.S. Army Logistics Innovation Agency Cost-Estimator Model

LIA has developed a QoL Cost Estimator model that computes the approximate capital cost of constructing certain kinds of facilities at contingency bases and the associated annual costs of logistics support required to maintain operations at those facilities. These estimates are based on engineering standards and technical specifications for shelters and buildings described in several Army and joint policy documents.[3] We compared the facilities found at our four case-study contingency bases with the facilities that the QoL Cost Estimator model described. We evaluated facilities on how closely they matched on the basis of their purpose, primary construction material, and total square footage. We focused on the December 4, 2012, reporting period when the QoL square footage peaked overall for these four bases.

We compared facilities[4] at our four case-study contingency bases with the LIA QoL Cost Estimator model and found that housing or billeting structures aligned better than other QoL facility types, primarily because, as a rule, housing dominated square footage for each base. As shown in Figure A.8, the percentages of aligned QoL-related housing square feet between our case-study bases and the QoL Cost Estimator model were 39 percent for Kandahar, 54 percent for Spin Boldak, 31 percent for Leatherneck, and 43 percent for Lindsey. After housing, latrines and showers vied with DFACs with respect to total square footage that either matched or did not match the QoL Cost Estimator model. All other QoL-related facilities ranked much lower in total square footage at these four bases and constituted only a few percentage points of the total QoL-facility square footage overall.

How facilities of contingency bases in southern Afghanistan align with the QoL Cost Estimator model suggests several things. First, facilities at these contingency bases vary widely in size and construction material, making them relatively nonstandard. Second, the overwhelming QoL-related spatial areas taken up at these four bases were dominated by housing

---

[3]   Primary references include HQDA, 2013a; AAFES Retail Contingency Matrix, June 11, Center of Standardization—Contingency Design Standards; AAFES, 2009; Applegate, 2013; USACE, 2013; Theater Construction Management System.

[4]   We categorized existing buildings and their purposes on four bases and compared them with the building types represented in the LIA model. The buildings matched if they were for the same purpose, within 10 percent of the LIA square-footage estimate, and constructed of similar materials (e.g., wood, concrete, or tent).

Figure A.8
Percentage of South Afghanistan Contingency Base Quality-of-Life Facilities and Space That Match Cost-Estimator Quality-of-Life Facilities

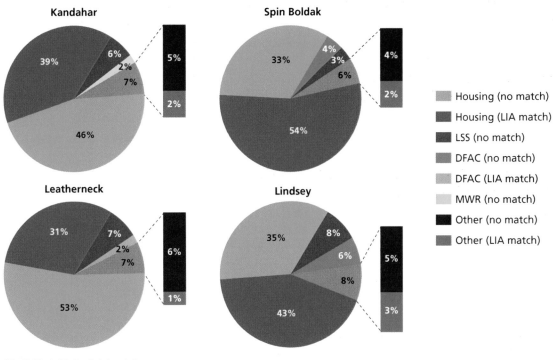

SOURCE: ACC–Rock Island data.
RAND RR1298-A.8

structures, followed by latrines and showers, DFACs, and MWR buildings. This is not surprising given the high value that personnel have for privacy and sleep, hot meals, and maintaining service and contractor hygienic conditions, fitness, and health. Next, we describe an analysis of facility types and power generation produced at Spin Boldak between May 2012 and March 2014, although not all data were available throughout this period.

Example of Facility Types and Power Generation: Spin Boldak

We analyzed facility types and power generation for Spin Boldak, one of our case studies. The population of Spin Boldak was primarily Army and contractor related; of the four contingency bases we analyzed in depth, it exhibited the greatest variation in population numbers over time. The data for facility types had greater detail for May 2012 to February 2013. Power-generation data were available for May 2012 to March 2014. We included all facility types for context and to account for all power generated at the contingency base. Some power generation was specific to particular buildings, while other power was generated in yards and available to multiple buildings and purposes. Where there was detail on building purpose, we categorized structures according to QoL-related purposes and combined non–QoL-related buildings or buildings that had missing data on their purpose as "other."

Most of the building space at Spin Boldak was related to billeting of service and contractor personnel, as shown in Figure A.9. The second-largest category of buildings is related to

Figure A.9
Spin Boldak Facility Types and Base-Supported Population Between May 2012 and February 2013

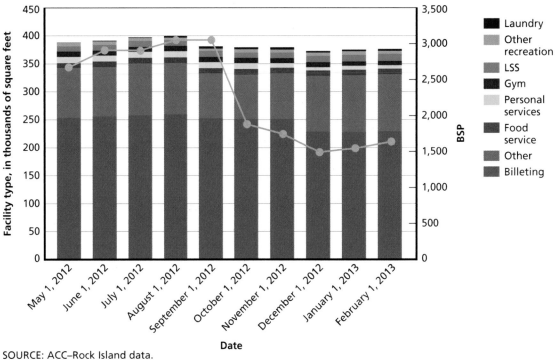

SOURCE: ACC–Rock Island data.
RAND RR1298-A.9

non-QoL purposes or had missing data and is labeled "other." All other QoL-related buildings, including food services, gyms, personal services, laundry, latrines, showers, and other recreation services, made up only 12 percent of the contingency base's building space. From May to September 2012, Spin Boldak's BSP was about 3,000 personnel. From October 2012 to February 2013, it decreased to half as many people, numbering about 1,500 personnel. Building square footage remained roughly constant over this period.

Next, we analyzed Spin Boldak's power generation by building purpose. For reasons described earlier, generators provided power generation. The generator data we analyzed pertained only to LOGCAP contractor reports. The Army purchased other generators into which we had no visibility. Some of the LOGCAP-reported generators were government-furnished property, and some the contractor purchased with government funds as contractor-acquired property.

Power-generation data were available for July 2012 to March 2014; however, more information on building purpose was available from the July 2012 to February 2013 MSOW data. As Figure A.10 shows, the single-largest use of power generation is associated with billeting and buildings that were missing information on their purpose. We surmise that most, if not all, of the "missing" data are associated with billeting. Over the analysis time period, the BSP was about 3,000 in July 2012 and decreased to about 1,000 by March 2014. (Figure A.9 shows a longer period of time for Spin Boldak's BSP data.)

We next computed correlations between BSP and power-generation demands of particular QoL-related building space—specifically, billeting, food services, water delivered, laundry, and personal services. Figure A.11 shows the trends. Spikes in power generation for billeting

Figure A.10
Spin Boldak Megawatt-Hours per Month and Base-Supported Population Between July 2012 and March 2014

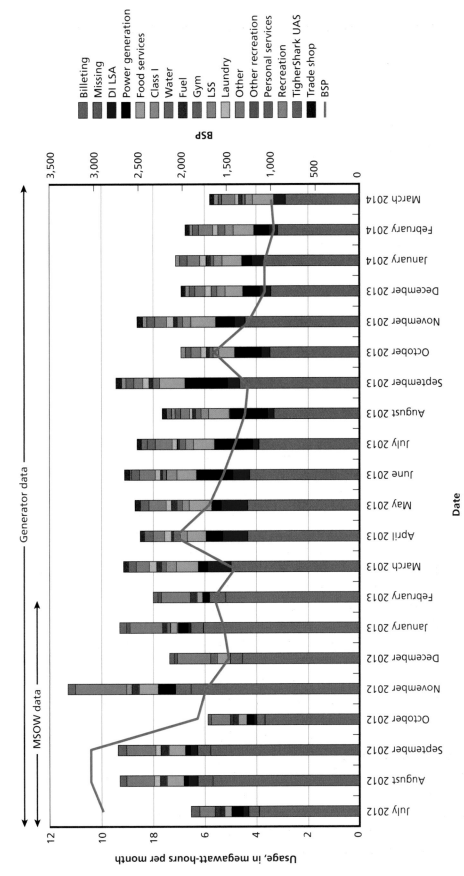

SOURCES: MSOW data, daily situation reports on power generation, power-generation status reports, and BSP reports.

RAND RR1298-A.10

Figure A.11
Spin Boldak Power-Generation Trends for Select Quality of Life–Related Building Space, July 2012 to March 2014

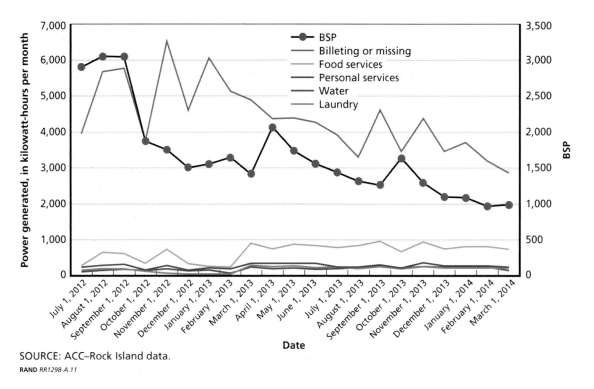

SOURCE: ACC–Rock Island data.

RAND *RR1298-A.11*

occurred in September, November, and January over the analytical time period. Power generation for food services and DFACs showed the second-highest demand for power, with other QoL-related services requiring much less power.

Correlations between Spin Boldak's BSP and the kilowatt-hours per month of several building types produced the results shown in Table A.3. The data showed that more power generation was needed as populations decreased. Water showed the highest negative correlation with food services, and even laundry was negatively correlated with BSPs. We hypothesize that food-service facilities generate nearly the same power- and water-consumption profile, regardless of base-camp population. The consumption rate per tenant will tend to increase

Table A.3
Correlations Between Spin Boldak Base-Supported Population and the Kilowatt-Hours per Month of Select Quality of Life–Related Buildings

Variable Pairs: BSP and	Correlation Value
Food services	−0.38
Water	−0.47
Laundry	−0.16
Personal services	0.10
Billeting	0.48

as camp population decreases. Also, BSPs might not accurately measure the total number of personnel accessing particular services that a base provides. For example, personnel located at smaller FOBs might not have certain services and rotate into larger bases on a temporary basis to have laundry done and eat hot meals.

Two variables that were positively correlated with BSPs were personal services and billeting. Personal services were weakly correlated with base populations, and billeting was moderately correlated. According to interviews of military RAND fellows who have previously been deployed, the QoL benefits associated with billeting are the ability to use electronics, computers, and Wi-Fi and having heating and air conditioning. Spikes in power generation for billeting that seemed to be out of phase with BSPs might be related to weather conditions that would increase demands for heating or air conditioning.[5]

Although the Army's preference is to have standard equipment, the number of power generators used in Afghanistan fell outside the normal equipment with which troops deploy and had to be purchased. When shortages of power generation occurred, generators were purchased from local vendors, which had a variety of generator brands and models. Tents and CHUs often had their own generators. As one person from Rock Island joked, every coffee maker had its own power generator. Statutory requirements constrain the Army and its contractors from developing systems of generators that could have been connected and controlled by "smart" regulators that would have made power generation more efficient, which meant that, all other things being equal, more generators were required to produce a given amount of power.

In February 2013, a period that had the most complete information for our four case-study sites, LOGCAP reports indicated that 54 percent of the 3,444 generators at all sites in southern Afghanistan were 100 kW or less, shown in Figure A.12.[6] In February 2013, Spin Boldak had 144 generators that were, for the most part, evenly distributed across generator capacities, with 37 percent having capacities of 100 kW or less.

Each of the bases we analyzed had a proliferation of generator brands, models, and capacities. Figure A.13 shows generators by make for sites in southern Afghanistan and Spin Boldak as of February 2013. Although two types of generators make up almost half of all generator types in southern Afghanistan and seven types of generators constituted 78 percent of the total number, there were 73 types of generators in all. Twenty types of generators had only one generator each across southern Afghanistan sites. The number of generator types was lower for Spin Boldak, although even it had 15 total types of generators. Two generators made up 46 percent of all generators at Spin Boldak in February 2013, and five made up 80 percent of the total generator types. Five generators had only one generator of that type on the base. Having a wide variety of generators creates higher maintenance and supply costs because more repair manuals and the wider variety of parts are required, which increases costs.

Next, we discuss fuel use at select bases as reported to the Tactical Fuels Manager Defense (FMD) data system.

---

[5] A study of energy consumption at Camp Buehring showed higher power-plant energy consumption during the summer months (i.e., June, July, and August) (Keysar, 2014).

[6] Power-generation data came from several files that were merged. They include the MSOW, situation reports daily on power generation, power-generation status reports, and BSP reports.

Figure A.12
Generator Capacities, All Sites in Southern Afghanistan and Spin Boldak, February 2013

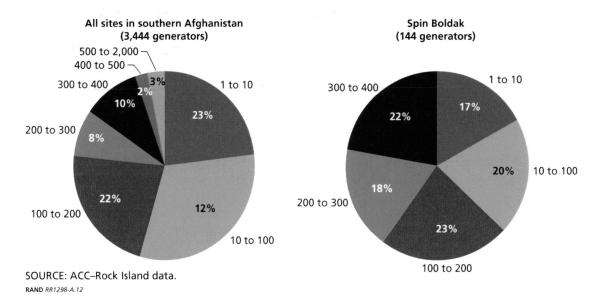

SOURCE: ACC–Rock Island data.
RAND *RR1298-A.12*

Figure A.13
Makes of Generators at All Sites in Southern Afghanistan and Spin Boldak in February 2013

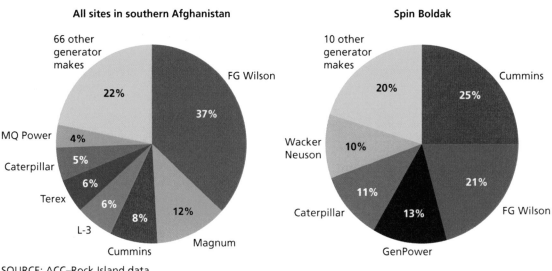

SOURCE: ACC–Rock Island data.
RAND *RR1298-A.13*

Tactical Fuels Manager Defense Data

The Tactical FMD program manager, under U.S. Army Materiel Command, provided the project team with data that report cumulative gallons of aviation gasoline; DF2; jet propulsion fuel, type 8; and MG1 vehicle fuel usage. The data cover 37 bases throughout Afghanistan from March 2011 through May 2014. The project team used the data to inform a background analysis of theater fuel usage by type and location.

Tactical FMD fuel is sorted into six major categories with 21 subcategories, as detailed in Table A.4.

Figure A.14 shows the proportion of average monthly fuel usage in the six major categories for four contingency bases in southern Afghanistan from June 2011 to May 2014, or three years. These FOBs had the most consistent reporting of fuel usage in the data set, which the Army Petroleum Center provided for its Tactical Fuels Manager Defense data system.[7] For three of the four bases, base energy ranked as the largest user of fuel. Leatherneck stands out as an anomaly because of its large non-Army fuel consumption. Most of the U.S. forces at the

Table A.4
Tactical Fuels Manager Defense Major Categories

Major Category	Definition
Aviation	All Army aviation
Installation energy	Base life-support fuel consumption
Non-Army	All non-Army customers, including other services, non-DoD, and non–U.S.-provided fuel without reimbursement. This category excludes total fuel captured in the "non-Army reimbursables" category.
Non-Army reimbursables	Non-Army customer data with significant fuel transactions for potential reimbursement through military interdepartmental purchase request
None	Unclassified equipment; awaiting accountable-office category designation
Tactical	All Army tactical equipment

Figure A.14
Average Monthly Fuel Used, by Activity, for Four Bases, June 2011 to May 2014

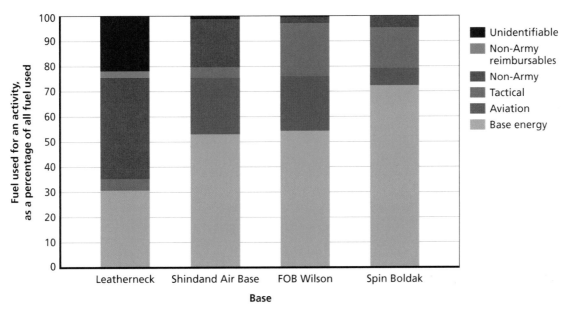

SOURCE: ACC–Rock Island data.

RAND RR1298-A.14

---

[7] Data were collected by measuring gallons consumed by purpose. Camp personnel self-reported usage. Of the 39 camps that collected data, only four recorded data throughout the three-year period of June 2011 to May 2014.

camp were U.S. marines. It is highly likely that much of the "non-Army" portion of use was for camp purposes. The second-highest category of fuel usage pertained to aviation and tactical consumption, which helicopters and ground vehicles used.

What Crowdsourcing Reveals About Quality-of-Life Demands at Contingency Bases

The Army conceives of contingency bases and QoL goods and services as falling into three groupings: basic, expanded, and enhanced (HQDA, 2013a, Chapter 1). Basic camps are "considered essential to sustain operations for a minimum of 60 days." They are built around a unit's organic capabilities and augmented with force-provider or similar systems (HQDA, 2013a, pp. 1-2, 1-3). "There is little to no contracted support affiliated with basic quality of life standards" (HQDA, 2013a, p. 1-4). Expanded camps feature

> capabilities that have been improved to increase efficiencies in the provision of base camp support and services, and expanded to sustain operations for a minimum of 180 days. . . . The expanded quality of life is intended to decrease the stress on personnel deployed for longer periods of time. (HQDA, 2013a, p. 1-4)

Enhanced camps have "expanded capabilities that have been improved to operate at optimal efficiency and support operations for an unspecified period of duration" (HQDA, 2013a, p. 1-3):

> Enhanced quality of life standards approach those of an installation. Normally these enhanced quality of life standards should not exceed those of a permanent base or installation, but because of the nature of the deployment, some support and services may need to. (HQDA, 2013a, p. 1-5)

As the research advanced, the project team realized that we needed the soldier perspective on QoL. DoD policies regarding QoL were clear, the engineering standards for contingency bases supporting QoL were well understood, and the MWR standards for delivering QoL goods and services were straightforward. The research team had also heard anecdotes from the Arroyo Center Army fellows but lacked a richer understanding of what *QoL* meant to soldiers and which QoL goods and services they most value while deployed.[1] A survey might have been the obvious way to capture soldier preferences but was beyond the scope of the project. As an affordable alternative, the research team turned to crowdsourcing (Brabham, 2013).

---

[1] These fellows are active and reserve component (RC) Army officers who serve yearlong fellowships in the Arroyo Center, the Army's federally funded research and development center for studies and analyses. See HQDA, 2012.

Adapting Crowdsourcing

Crowdsourcing has many applications. The research team adopted the "broadcast-search" approach, soliciting answers via email to a limited number of questions from soldiers with overseas deployment experience. The team asked the Army fellows and other Army acquaintances and colleagues for help. The research team provided text for the email, explaining the project and its objectives and asking the recipient to answer a few questions at a website connected to the RAND public internet home page. The email solicitation also asked the recipients to pass our request for assistance along to their associates who had deployed to contingency bases.

Our expectation was that soldiers would respond to the email questionnaire because (1) the subject of QoL on contingency bases was of interest to them and (2) because the request came via another soldier whom they respected. The research team would download the responses from the web page and analyze them. The questions themselves were simple and open ended:

- When and where were you deployed?
- What was your military occupational specialty and grade?
- What factors contributed to your QoL while deployed (e.g., hot meals, laundry service, running water)?
- What factors detracted from your QoL while deployed (e.g., lack of sleep, insufficient potable water)?
- If you could have added any one QoL element to your tour, what would it have been?

The answers to these questions should allow the research team to infer an operational definition for *QoL*. The answers should also allow us to identify the high-value goods and services and differentiate them from the lower-value goods and services available at overseas contingency bases.

The Respondents

The crowdsourcing email prompted 34 responses: initially disappointing. Among the respondents were 27 officers, two warrant officers, and five enlisted soldiers. The most senior respondent was a major, the most senior warrant officer a chief warrant officer 3, and the most senior noncommissioned officer was a first sergeant. These soldiers had vast experience with contingency bases, reporting service in 72 different countries and regions, summarized in Table B.1.

The respondents also had an extensive deployment history, illustrated in Figure B.1. This history spanned the years 1990 through 2014 and included a total of 110 years deployed.

The respondents reported 36 different military occupational specialties and functional areas. The respondents tended to be members of the combat support and CSS communities, leaving the combat arms groups underrepresented, although cavalry, infantry, field artillery, and special forces were among the respondents.

Clearly, crowdsourcing did not deliver a statistically significant sample of soldiers whose views on base-camp QoL the study could generalize. The sample is better thought of as a group of subject-matter experts, a group that has experienced base-camp QoL over the course of 24 years through 110 deployments at 72 different locations.

Table B.1
Deployment Locations Reported by Respondents

Location	Respondents Reporting Duty There
Afghanistan	29
Iraq	28
Kuwait	3
Saudi Arabia	2
Kosovo	2
Albania	1
Bosnia	1
Djibouti	1
Former Yugoslav Republic of Macedonia	1
Panama	1
Philippines	1
Qatar	1
Sinai	1
Total	72

Figure B.1
Respondent Deployment History

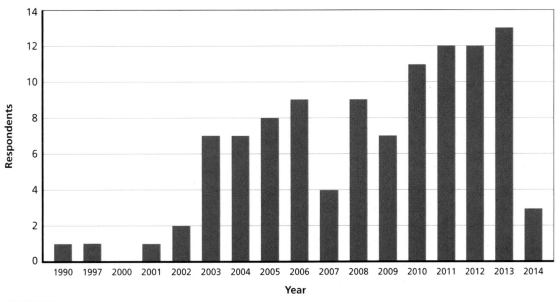

RAND *RR1298-B.1*

Responses

This section describes the respondents' answers to the questionnaire. It proceeds question by question.

What Factors Contributed to Your Quality of Life While Deployed?

This open-ended question was intended to help the research team understand, from a soldier perspective, what exactly constituted QoL (see Table B.2). The most frequently mentioned things would represent the substance of QoL for deployed soldiers. Some of the responses are free of ambiguity (e.g., hot meals). Other answers (e.g., running water) are surrounded by it. Running water for what—flush toilets, showers, laundry? In instances in which answers were clearly dealing with a single good or service but used different words to describe it, we have

Table B.2
Respondents' Quality-of-Life Contributors

Contributor	Number of Mentions
Hot meals	25
Laundry service	22
Internet access, communications home	19
Gym	15
Running water	15
Showers	11
MWR, movies, shopping, entertainment, diversion	10
Climate control	5
Privacy or restful sleeping space	5
AAFES or PX	4
Good DFAC	4
Alternative food source	3
Electricity	3
Leadership, rational policies	3
Mail	3
Hardened building	2
Accountable contractors	1
Bottled water	1
Fresh fruit	1
Group exercise options	1
Proximity to facilities	1
Running space	1
Total	155

bundled those responses into a single term. In instances in which a term is vague (e.g., running water), we have not forced it into a larger category of responses, leaving it to stand on its own merits.

Some of these responses deserve some explanation. *Internet access, communications home* is a term representing a bundle of responses that referenced internet access and some ability to communicate home. This bundle of responses might reflect a desire for easier access and more privacy when using communications home. We infer this context because morale satellite (morale sat) has become widespread, and MWR facilities make internet available. It might be, therefore, that soldiers prefer not to wait in lines for their turn on the network and want greater privacy when connected.

Privacy or restful sleeping space represents a coupling of desires that respondents reported. The desire for privacy seems widespread. In addition to its appearance in the crowdsourcing responses, privacy was also a topic in our discussions with the Arroyo Center Army fellows. Both leaders and subordinates need time away from each other. Restful sleeping space might be linked with privacy because respondents want private quarters. A desire for restful sleeping space might reflect the challenges of getting restful sleep because of ubiquitous generator noise, the inability to shut out daylight, the noise and activities of other soldiers sharing the same quarters, and a lack of climate control where heat or cold makes restful sleep difficult.

Leadership, rational policies might refer to a desire for relaxation of some of the stateside rules that soldiers find unnecessary and demeaning at overseas locations. Some of the respondents said as much. *Rational policies* might also be code for the end of the prohibition of consumption of alcoholic beverages.[2] *Leadership* might also mean having senior officers who insist that contractors perform the functions they are contractually obligated to perform.

Accountable contractors has to do with contractors that meet their responsibilities for the maintenance, functionality, and cleanliness of camp facilities (perhaps especially latrines).

What Factors Detracted from Your Quality of Life While Deployed?

Respondents cited 155 factors that added to their QoL. In this section, they identify 97 detractors. The detractions from QoL are more disparate than the factors that enhanced QoL. For example, seven out of 22 categories of enhancements to QoL had ten or more respondents. For the detractors, only one category, sleep, received more than ten responses. Table B.3 summarizes the responses.

If You Could Have Added Any One Quality-of-Life Element to Your Tour, What Would It Have Been?

The final question asks respondents for the one QoL element that would have made their deployment better. Eight respondents replied "nothing" or did not answer the question. Several respondents indicated that deployments are supposed to be austere. Others offered the view that job satisfaction should be enough to make conditions tolerable. Table B.4 summarizes the responses.

Routine Skype seems to go beyond a desire for easy access to regular communications home. Skype provides real-time video and audio, which affords soldiers the opportunity to

---

[2] Research team members have heard comments to the effect, "You trust me with a security clearance and a loaded weapon but not with a couple of beers?!"

Table B.3
Respondents' Quality-of-Life Detractors

Detractor	Number of Mentions
Lack of sleep	17
Unhygienic conditions	8
Extreme temperatures	6
OPTEMPO	5
Poor leadership	5
DFAC shortcomings	4
Lack of communications home	4
Lack of laundry service	4
Lack of showers	4
Unnecessary rules	4
Boredom, no MWR	3
Insufficient potable water	3
Lack of privacy	3
Lack of running water	3
Bad facilities	2
Electrical issues	2
Lack of personal freedom	2
Population disproportionate to the available facilities	2
Distance from security forces	1
Inequitable duties with other units	1
Infrequent fresh fruit	1
Infrequent prepared meals	1
Insects	1
Insufficient physical training	1
Irregular mail	1
Lack of fitness facilities	1
Lack of interpreters	1
Mistreatment of third-country nationals	1
Mud	1
No beer	1
Non-DoD civilian interference in mission planning	1
Poor contractor accountability	1
Porta-potties	1

Table B.3—Continued

Detractor	Number of Mentions
Redundant work	1
Total	97

Table B.4
Other Desirable Quality-of-Life Features

Feature	Number of Mentions
Routine Skype	10
Nothing	8
Better sleeping conditions	4
Quiet, solitary, private	4
Free internet in rooms	3
Healthy food options	2
Regular access to entertainment	2
24-hour food service	1
Air conditioning and heat control	1
Beer ration	1
Flush toilets	1
Greater access to in-country R&R	1
Improved capability for hygiene	1
Integrated bathroom and quarters	1
More-developed quarters, amenities	1
More liberties	1
More showers	1
Quality of leadership	1
Running water	1
Therapy dogs	1
Total	46

interact with their families. One respondent suggested that it would enable him to watch the birth of his next child (assuming that the requisite privacy was available).

Analysis of the Crowdsourcing Data

The analysis of the crowdsourcing data is intended to answer two questions. First, how did the respondents understand QoL? The most frequently mentioned items should lie at the heart of the resulting operational definition. Second, the analysis should identify options for the Army to satisfy the demand for goods and services that appear as part of the operational definition of *QoL*.

Defining Quality of Life from Crowdsourcing

The crowdsourcing replies contain elements of things that respondents indicated improve their QoL when deployed, things they indicate detract from their QoL while deployed, and things they would very much like to have while deployed. Many of these items, however, received single or very small numbers of mentions in the crowdsourcing. Therefore, to avoid including elements that received very few mentions, the research team determined to bound the definition of *QoL* by including the top approximately 50 percent of responses from each category.[3]

Thus, in terms of things respondents reported most valuing, 16 percent reported hot meals, 14 percent reported laundry, 12 percent reported internet and communications home, and 10 percent reported the gym.

In the case of things respondents reported detracted from their QoL while deployed, 18 percent reported lack of sleep, 8 percent unhygienic conditions, 6 percent extreme temperatures, 5 percent poor leadership and high OPTEMPO, and 4 percent reported lack of communications home and unnecessary rules. The analysis does not treat poor leadership or high OPTEMPO further because neither of these conditions is responsive to actions the G-4 and sustainment community could take.

In the final category, things respondents reported as highly desirable, 22 percent reported routine Skype; 17 percent reported a desire for quiet, solitude, privacy, and better sleeping arrangements; and another 17 percent answered "nothing" or did not answer the question. Combining some categories carefully and turning negative detractors into positive attributes, the operational definition of *QoL* emerging from the crowdsourcing can be summarized as follows:

- Provide hot meals.
- Make laundry service available.
- Create conditions for quality sleep.
- Maintain hygiene standards.
- Provide a gym.
- Provide Skype and private communications home.
- Provide privacy, solitude, and sleeping arrangements.

Options for Satisfying Demand for Quality of Life

The options for satisfying the demand for QoL goods and services will vary with the type of deployment and the phase of operations.[4] Phase 0 represents peacetime activities and is characterized as shaping activities. Phase I is the deterrence phase, Phase II the shaping-operations phase, Phase III the dominate phase, Phase IV the stabilize phase, and Phase V the enabling-civil-authority phase. The discussion in this section does not address Phase 0 because deployments in this phase are generally peacetime, noncombat, and of relatively short dura-

---

[3]   In other words, we sum the percentages of respondents reporting a certain response until the sum is equal to or greater than 50 percent.

[4]   According to Joint Chiefs of Staff, 2011, Figure V-3, p. V-6.

tion; Phase V receives no discussion because the QoL dimension of this phase will reflect force-withdrawal policies and will be largely residual from Phase IV. Table B.5 summarizes the main features of QoL delivery in Phases I through IV.

The conditions summarized in Table B.5 will vary with local circumstances. The Phase I description of meal service is typical of food service at Camp Wolf, Kuwait, during the deterrence phase before the U.S. invasion of Iraq in 2003 and a vast improvement over conditions during Operation Desert Shield in 1990.

Contractors can provide laundry service during Phase I because deterrence operations typically take place on the territory of a friendly state (e.g., Kuwait). The Army has not had to face protracted cases of Phase II and Phase III operations, but, if it did, the CSS community would have to provide field bath and clothing DX facilities to sustain combat operations.

Sleeping conditions are difficult to satisfy, although a policy of noise and light discipline in sleeping spaces, segregation of sleeping areas from other duty areas, and temperature control could make a significant improvement.

Hygiene is demanding. The current standard is a compromise between aesthetic considerations and operations and maintenance costs. Soldiers quickly learn the cleaning schedules for their facilities and take steps to use them immediately after they have been refreshed and replenished with expendables.[5]

Privacy and solitude improve as the tempo of operations declines, in-theater population shrinks, the number of operational contingency bases contracts, and privacy becomes easier to accommodate. It might be realized at different points in time at different contingency bases as the local population ebbs.

Conclusions and Recommendations

The crowdsourcing definition of *QoL* can be understood through at least two distinct lenses: first, through the lens of base-camp engineering standards, and second, through the lens of

Table B.5
Quality-of-Life Delivery in Phases I Through IV

Good or Service	Phase I	Phases II and III	Phase IV
Meals	DFAC serves fresh meals	Five days of supply of MREs	DFAC serves four meals/day
Laundry	Drop-off/pickup	CSS-operated, field-operated bath point and clothing DX	Initial DX followed by on-site laundry service
Sleep conditions	Temperature controlled with noise, light discipline	Undersatisfied	Temperature controlled with noise, light discipline
Hygiene	Repurposed space plus temporary facilities	CSS-operated field bath point or "baby-wipe" showers	Showers, latrines 1:20 ratio
Communications home	Mail, commercial internet, morale sat	Mail	Morale sat, MWR, soldier-contracted services
Privacy and solitude	Undersatisfied	Undersatisfied	B-hut or equivalent

---

[5] Interview with LOGCAP program management, June 22, 2014.

Maslow's hierarchy of needs (Maslow, 1943). The base engineering standards establish the design criteria for camps throughout their life cycles, from initial arrival until they are destroyed or transferred to local officials.[6]

Base-Camp Engineering Standards and Quality of Life

Base-camp engineering standards establish the design criteria and evolutionary patterns of camps through their life cycles. There is a deterministic character to the standards in that, if nothing about the situation changes beyond the passage of time, a camp will almost certainly evolve to provide a better living and working environment driven by design standards that are, in part, time-based.

In Figure B.2, we have coded the different types of QoL goods and services the crowd-sourcing respondents indicated they value to reflect their association with basic, expanded, and enhanced base-camp standards. The blue bracket represents 80 percent of all the QoL elements found in the figure. Note that the basic level of base-camp services can satisfy most of these elements. One would need only to cherry pick a few elements from the expanded and enhanced categories (e.g., Skype, privacy, solitude) to satisfy much of the demand identified in the crowdsourcing.

Figure B.2
Quality of Life and Base-Camp Standards

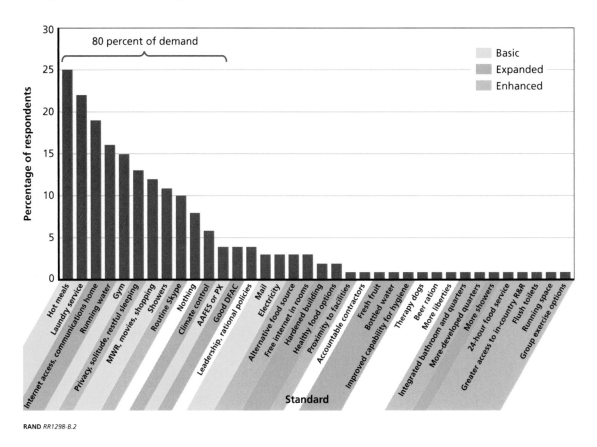

RAND *RR1298-B.2*

---

[6]   See, for example HQDA, 2010a. Other documents containing engineering standards are listed in the references at the end of this report.

Maslow's Hierarchy of Needs and Quality of Life

Figure B.3 shows the QoL elements coded against the categories within Maslow's hierarchy of needs (Kerce, 1992). In this case, by satisfying 78 percent of the demand for QoL as determined from the crowdsourcing, the Army would be able to address all of the categories in Maslow's hierarchy, though some would receive only minor treatment. This observation suggests that much QoL demand might be satisfied—in terms both of base-camp development and Maslow's hierarchy—at relatively modest levels of investment.

Recommendations

Verify the results from crowdsourcing. To ensure consistency and fidelity among results when surveying soldiers with deployment experience, asking the same questions posed in the crowdsourcing effort would be best. The U.S. Military Academy's social sciences department could be tasked to develop the sampling and collection strategy and to implement the survey. It might be a project suitable for cadets during their summer training.

Standardize QoL packages. Design each package to be responsive to an ever-greater percentage of soldier demand for QoL goods and services. For example, the basic standardized package might address the 80-percent solution that our analysis suggests. The expanded model might satisfy demand to the 90th percentile, while the enhanced form would satisfy 100 percent of reported demand.

Figure B.3
Maslow's Hierarchy of Needs and Quality of Life

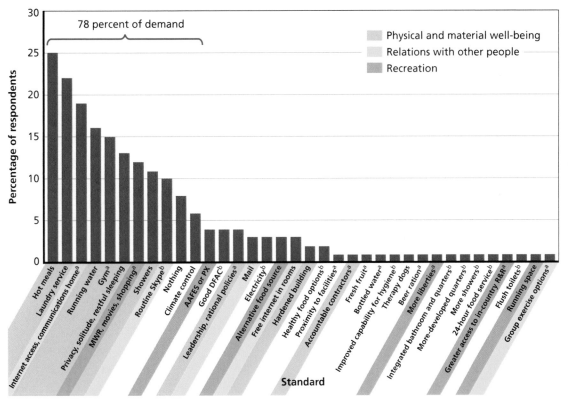

[a]Personal development and fulfillment.
[b]Social, community, and civic activities.

Of course, such an approach responds only to the demand identified through crowd-sourcing, and other considerations bear on base-camp design and operation. High among these considerations lies efficiency, especially in terms of fuel and water consumption. Cost and administrative considerations influence the range of operational power solutions available at contingency bases and often result in very inefficient use of generators—sometimes only 18 percent of their available capacity ("History of LOGCAP in Afghanistan," 2014).

Water is the second major commodity that is not cost-effectively procured, delivered, consumed, and recovered. Policy in Afghanistan requires the Army to drill down to the second-tier aquifers to reach water, often 1,500 feet below the surface. The practice leaves the water in upper aquifers available for local agriculture and in reach of its drilling technology ("History of LOGCAP in Afghanistan," 2014). Water must be trucked to U.S. contingency bases (some 130 camps in Afghanistan) and gray and black water must be transported out for recovery and treatment. The more efficient the drilling and well operations, the more efficient the delivery; the more efficient the recovery and processing, the less water will cost per gallon.

Note that the basic level of base-camp services can satisfy most of these elements. One would need only to cherry pick a few elements from the expanded and enhanced categories (e.g., Skype, privacy, solitude) to satisfy much of the demand identified in the crowdsourcing.

Restorative Effects of Quality-of-Life Goods and Services

Introduction

To determine what factors contribute to soldiers' well-being and how these factors can be implemented to mitigate or eliminate stressors unique to being deployed and stationed abroad at locations in or near combat theater, we reviewed six pieces of literature. As part of the review, we undertook three key tasks:

1. Identify what factors contribute to soldiers' well-being in general, including QoL, resilience, and fitness.
2. Identify stressors unique to soldiers stationed abroad at temporary bases in addition to those common to all military personnel.
3. Determine methods and paths to implementation to mitigate stressors and improve overall QoL, resilience, and well-being of soldiers stationed at temporary bases abroad.

The subsequent sections address each in order.

Factors That Contribute to Soldier Well-Being

The scope of this literature review includes identifying which factors contribute to soldier well-being, with a special focus on personnel stationed at temporary bases abroad. It is clear that several components contribute significantly to overall soldier well-being, including QoL, resilience, and fitness (medical, psychological, and physical). The latter two are the focuses of a series of eight reports that RAND released between 2013 and 2015, each one focusing on a fitness domain and applicability to airman fitness.[1] Of these eight fitness domains, we examine three herein: psychological, physical, and medical. We review soldier well-being in terms of QoL, resilience, and fitness. At the end of the appendix, we provide summaries of relevant constructs.

Quality of Life

Soldier QoL is seen as a contributor to soldier well-being: Army programs focusing on improving QoL are supposed to enhance QoL and thus well-being (Sims et al., 2013). QoL is gen-

---

[1] See Flórez, Shih, and Martin, 2014; McGene, 2013; Robson, 2013, 2014; Robson and Salcedo, 2014; Shih, Meadows, and Martin, 2013; Shih, Meadows, Mendeloff, et al., 2015; Yeung and Martin, 2013; and Meadows, Miller, and Robson, 2015.

erally perceived as important to the Army's goals of recruitment, retention, and readiness, though formal research establishing strong ties between QoL and these goals has yet to occur. The connection between QoL and retention and motivation is not unique to the Army: The Navy has recognized that "[a]n all-volunteer U.S. military force has emphasized the need to [ensure] that the life quality experienced by military members and their families will attract, motivate, and retain qualified personnel" (Kerce, 1992, p. 1). Kerce defines *QoL* as "the degree to which the experience of an individual's life satisfies that individual's wants and needs, both physical and psychological" (Kerce, 1992, p. vii). Although this definition is older, it gives insight into the role that different components of physical and psychological resilience play into QoL (see the next section, "Fitness and Resilience"). However, a formal, unified definition of QoL has not been agreed on across Army programs (Sims et al., 2013), a fact that hinders proper evaluation of QoL and how it relates to well-being. Promoting a "research road map" in this area should be a priority for the Army (Sims et al., 2013). What is clear from the literature is that QoL encompasses multiple domains; according to Sims et al., 2013, the *Report of the 1st Quadrennial Quality of Life Review* (Office of the Under Secretary of Defense for Personnel and Readiness, 2004) described QoL as being made up of the 14 domains (Sims et al., 2013). However, most of the domains listed, such as child care, do not apply to QoL on contingency bases.

Although the domains listed in *Report of the 1st Quadrennial Quality of Life Review* might be adequate to describe QoL for soldiers who are garrisoned, they are insufficient for describing QoL on contingency bases. Flanagan, 1978, as Kerce notes, describes five different classes of domains that we utilize in our research (Kerce, 1992, p. 10):

- *Physical and material well-being* includes the domains material well-being and financial security and health and personal safety.
- *Relations with other people* includes the domains of relations with spouse, having and raising children, relations with other relatives, and relations with friends.
- *Social, community, and civic activities* includes the domains of activities related to helping or encouraging others and activities related to local and national government.
- *Personal development and fulfillment* includes the domains intellectual development, personal understanding and planning, occupational role (job), and creativity and personal experience.
- *Recreation* includes the domains socializing, passive or observational recreation activities, and active or participatory recreational activities.

These five classes of domains are broader based and are better suited to meet the qualifications of base-camp QoL evaluation.

For example, when a soldier is stationed at a base abroad, MWR opportunities might be of greater immediate importance to that soldier than tuition assistance. This theme is described in depth—using examples—in work that Marlowe, 2001, presents for evaluating stressors and the well-being of soldiers stationed at temporary bases abroad during Operation Desert Storm.

Finally, *Commander and Staff Officer Guide*, Army Tactics, Techniques, and Procedures (ATTP) 5-0.1, provides logical categories for QoL support and services (HQDA, 2011).

Annex F is the format for the sustainment annex of the base OPORD. Within Annex F, the sustainment paragraph consists of four major subparagraphs:

- Material and services includes classes of supply, field services, transportation and distribution, maintenance, contract support, and contract labor.
- Personnel includes morale development and support, headquarters management, and strength maintenance.
- Health-system support includes preventive medicine, medical and dental services, and veterinary services.
- Host-nation support is requested as required.

These subparagraphs capture the vast majority of QoL-related commodities and services. In particular, the commanders' management of materiel and services and personnel will create the largest QoL impact on the base camp. Table C.1 outlines examples of QoL elements in the material and services and personnel categories.

Fitness and Resilience

Fitness and resilience are said to be major components of well-being. *Resilience* is defined as "the ability to withstand, recover from, and grow in the face of stressors," while *fitness* is defined as "a state of adaptation in balance with the conditions at hand" (Shih, Meadows, and Martin, 2013, p. 6; originally from Mullen, 2010). Shih, Meadows, and Martin, 2013, describes the connection between resilience, fitness, and well-being in a two-model framework that Meadows, Miller, and Robson, 2015, developed, in which the first model describes the direct effects that resilience can have on well-being and the second model describes its indirect effects on well-being (Shih, Meadows, and Martin, 2013, p. 6). On the model of main (or direct) effect of resilience of well-being, they wrote,

> [The association between resilience and well-being] does not depend on whether a stressor is present. Experiencing a stressor also has an independent, direct effect on mental health . . . .[2] In this model, it is not necessary to know whether an individual has experienced a stressor to assess whether the resilience factor has an effect on well-being.

Table C.1
Quality-of-Life Elements, by Category

Category	Element
Material and services	Housing, meals and feeding, shower, latrine, and waste removal and recycling
Personnel	Chapel and religious services, internet café, fitness center, AAFES, movie theater, telephone call center, educational center, other vendor services (including local bazaar, post office, and banking), USO, and base-camp transit or bus route

---

[2] And thus well-being.

And on the model of buffering (or indirect) effect of resilience on well-being:

> In this model, it is necessary to know whether an individual has experienced a stressor to assess whether resilience and resilience factors have an effect on well-being. That is, resilience can be understood only in the context of stress. Most resilience resources have a direct effect on well-being, contribute to an individual's overall fitness level, *and* can be used to combat stress or strain when it occurs.

Although Shih, Meadows, and Martin, 2013, focuses on medical resilience specifically, the models are derived from the 2015 Meadows, Miller, and Robson report on resilience in general, and one can consequently assume that this definition of *resilience* is a general one. As part of the literature review conducted, we considered three primary categories of fitness and resilience, and we discuss each in depth in the rest of this section.

Psychological Fitness and Resilience

Psychological fitness and resilience are found to be crucial to well-being by furnishing one with the ability to deal with stress in an effective manner. We considered two reports that addressed theoretical frameworks of psychological fitness and resilience: *Psychological Fitness and Resilience: A Review of Relevant Constructs, Measures, and Links to Well-Being* (Robson, 2014), and "Psychological Fitness" (Bates et al., 2010).

The Robson report presents an extensive catalog of psychological fitness constructs, recommendations for enhancing these skills (or an overview of research regarding these enhancements), and applicable measurements. Here we enumerate these constructs, applicable definitions, and how they relate to well-being:

- self-regulation (from Karoly, 1993): "[S]elf-regulation refers to those processes internal and/or transactional that enable an individual to guide his/her goal-directed activities over time and across circumstances" (Robson, 2014, p. 10). Self-regulation "facilitate[s] the ability to exercise restraint, direct choices, and persist in the face of adversity [and] is also important in helping individuals to 'bounce back' from stress" (Robson, 2014, p. 10).
- coping strategies (from Skinner et al., 2003): "Coping involves the ways people actually respond to stress, such as through seeking help, rumination, problem solving, denial, or cognitive restructuring" (Robson, 2014, p. 11).
- positive and negative affect (partially taken from Carver and Harmon-Jones, 2009): "Affect generally refers to an individual's subjective sense of positivity or negativity arising from an event" (Robson, 2014, p. 14). Robson notes that there are significant (positively correlated) connections between negative affect and negative effects on well-being and positive affect and enhancing well-being.
- perceived control: "Perceived control can be defined by the extent to which people feel a sense of control over events (i.e., locus of control) as well as being the initiator of their own behavior" (Robson, 2014, p. 15). There are two instances of locus of control: Those with *internal* loci of control believe that they have control over their lives and events, while those with *external* loci of control attribute causes of events to others, chance, or fate. Robson notes that locus of control has been associated with positive work outcomes, the ability to deal with stress more readily, engaging in task-focused coping behaviors,

additional motivation, emotional well-being, higher self-esteem, and better overall health (Robson, 2014, p. 16; includes components from multiple authors).

- self-efficacy (partially taken from Bandura, 1982): Self-efficacy is "concerned with judgments of how well one can execute courses of action required to deal with prospective situations" (Robson, 2014, p. 17). Self-efficacy is positively associated with job performance and satisfaction, academic performance, less stress, and potentially enhancing the immune system (by building efficacy in coping with stress) (Robson, 2014, p. 17).
- self-esteem (adapted from Maslow, 1943): "Self-esteem, concerned with the global evaluation of one's self-worth, is an important marker for overall well-being . . . . Self-esteem is associated positively with satisfaction and subjective well-being and supporting happiness," though "the data are not sufficient to justify development of efforts simply to raise self-esteem (Robson, 2014, [p.] 18)."
- optimism (partially taken from Prati and Pietrantoni, 2009): "Dispositional optimism has been defined as the 'generalized expectancy for positive outcomes'" (Robson, 2014, p. 19). Optimism is shown to have clear benefits for well-being. However, optimism "may also present certain drawbacks and risks. For example, high levels of optimism may be maladaptive when facing persistent stressors. Nonetheless, the clear majority of research shows that optimism produce[s] positive effects and pessimism yields negative outcomes" (Robson, 2014, p. 20).

In addition to the seven psychological constructs enumerated above, Robson discusses several resources denoted as emerging constructs. Because Robson denotes these constructs as emerging, they do not have significant bases of research shown in the literature. These concepts continue the list above.

Adaptability

> Closely related to constructs of coping, adaptability involves how an individual or team adjusts in response to novel or changing environments [and] can aid not only in responding to stress but also in adjusting to new life roles (e.g., parenting, marriage), work roles (e.g., promotion, deployment), and evolving job demands (e.g., technology, new supervisor). (Robson, 2014, p. 21)

Although adaptability is intuitively supportive of resilience, clear definitions and measurements of adaptability do not yet exist in the literature.

Self-Awareness

According to Robson, "Military personnel often refer to self-awareness, or the historical maxim, 'Know Thyself' proffered by Plato and Socrates, simply as the 'Gut Check'" (Robson, 2014, p. 22). Some research supports a positively correlated connection between self-awareness and building resilience (as a precursor) and between self-esteem and mental health (introspection) and self-awareness as a protective factor for anxiety and depression (Robson, 2014, p. 22).

Mindfulness

According to Brown and Ryan, 2003, "Mindfulness 'is most commonly defined as the state of being attentive to and aware of what is taking place in the present'" (Robson, 2014, p. 23). Emerging research shows that mindfulness "may have a significant role [in] adaptive decisionmaking," is pertinent to well-being, and "can reduce mood disturbance and stress in cancer patients" (Robson, 2014, p. 13).

Table C.2 shows recommendations drawn from research summaries in the Robson report.

In addition to the Robson report, the Bates et al., 2010, article presents a military demand–resource model for psychological fitness, which the authors define as "the integration and optimization of mental, emotional, and behavioral abilities and capacities to optimize performance and strengthen the resilience of warfighters" (Bates et al., 2010, abstract). The authors additionally note the connection between psychological resilience and overall soldier well-being: "[equal] attention to the psychological component is critical for achieving the mind-body balance as desired in a total force fitness framework for military forces today" (Bates et al., 2010,

Table C.2
Summary of Soldier Well-Being Constructs

Well-Being Component	Source	Description
QoL	Sims et al., 2013; Kerce, 1992	
Family support	Sims et al., 2013, from Office of the Under Secretary of Defense for Personnel and Readiness, 2004	QoL is "the degree to which the experience of an individual's life satisfies that individual's wants and needs, both physical and psychological" (Kerce, 1992, p. vii).
Counseling services		
Financial planning		
Housing		
Child care		
Military spouse employment and career opportunities		
DoD schools		
Educational transitions among military children		
Commissary and military exchange systems		
Support for victims of domestic violence		
Support during the deployment cycle		
MWR opportunities		
Tuition assistance for voluntary education		
Partnerships with the states		
Physical and material well-being	Kerce, 1992, from Flanagan, 1978	
Relations with other people		
Social, community, and civic activities		
Personal development and fulfillment		
Recreation		

Table C.2—Continued

Well-Being Component	Source	Description
Fitness and resilience		
Psychological fitness and resilience	Robson, 2014; Bates et al., 2010	Resilience is defined as "the ability to withstand, recover from, and grow in
Self-regulation	Robson, 2014	the face of stressors," while fitness is defined as "a state of adaptation in
Coping strategies		balance with the conditions at hand" (Shih, Meadows, and Martin, 2013,
Positive and negative affect		p. 6; originally from Mullen, 2010).
Perceived control		
Self-efficacy		
Self-esteem		
Optimism		
Adaptability		
Self-awareness		
Mindfulness		
Awareness	Bates et al., 2010	
Beliefs and appraisals		
Coping		
Decisionmaking		
Engagement		
Physical fitness and resilience	Robson, 2013	
Medical fitness and resilience	Shih, Meadows, and Martin, 2013	

abstract). The military demand–resource model presented includes the desired psychological state (or demand) as output, coupled with two inputs, internal and external resources. External resources include "leadership, unit members, families, educational and training programs, and community support organizations and programs" (Bates et al., 2010, p. 22). In this context, QoL can be considered an external resource because the examples of external resources explicitly include different dimensions of QoL (families, educational and training programs, community support organizations). Internal resources are categorized into five psychological constructs, some of which the Robson report also discusses (includes definitions taken from the report):

- awareness: "Self-awareness is broadly defined as the self-descriptions that a person ascribes to oneself that influence one's actual behavior, motivation to initiate or disrupt activities, and feelings about oneself" (Bates et al., 2010, p. 28).
- beliefs and appraisals: "Beliefs are defined as psychological states in which an individual holds a premise to be true. . . . Appraisals have been theoretically linked with responses to stress and performance outcomes" (Bates et al., 2010, p. 29).

- coping: "Coping can be broadly defined as thoughts and behaviors a person uses to manage the demands of stress and to maintain optimal levels of energy and capacity to work" (Bates et al., 2010, p. 30).
- decisionmaking: "Decision making is defined as thoughts and behaviors used for evaluating and choosing courses of action to solve a problem or reach a goal" (Bates et al., 2010, p. 31).
- engagement: "Engagement is a sustained experience of strong identification with unit members, unit, and mission characterized by high levels of energy and full involvement in mission tasks. Engagement is characterized by three factors: dedication, vigor, and flow" (Bates et al., 2010, p. 31).

Table C.2 summarizes recommendations on increasing psychological fitness and resilience and, consequently, soldier well-being presented in the literature.

Physical Fitness and Resilience

Similarly to psychological resilience, physical fitness, exercise, and activity are found to be key components of overall well-being. Physical fitness "is defined as a set of health or performance-related attributes relating to the activities and condition of the body" (Robson, 2013, p. ix). The most direct benefits of increased physical fitness and activity are strong components of well-being:

> Strong evidence demonstrates that more active men and women have lower rates of all-cause mortality, coronary heart disease, high blood pressure, stroke, type 2 diabetes, metabolic syndrome, colon cancer, breast cancer, and depression. Strong evidence also supports the conclusion that more physically active adults and older adults exhibit a higher level of cardiorespiratory and muscular fitness, have a healthier body mass and composition, and have a biomarker profile that is more favorable for preventing cardiovascular disease and type 2 diabetes and for enhancing bone health. Modest evidence indicates that physically active adults and older adults have better sleep and health-related quality of life. (Robson, 2013, p. 13)

Increased physical fitness or exercise is additionally linked with increased ability to cope with stress and can "protect against the onset of certain mental health disorders and symptoms, including depression and anxiety" (Robson, 2013, p. 14). However, physical exercise has some potential drawbacks, especially applicable to soldiers stationed at temporary bases abroad. These potential drawbacks act as stressors, and we discuss them in "Stressors and Detriments to Well-Being."

Medical Fitness and Resilience

Medical fitness, in a definition that is partially taken from "the special issue of *Military Medicine* on Total Force Fitness" (Shih, Meadows, and Martin, 2013, p. 5) and Mullen, 2010, "for Airmen, their families, and Air Force civilians entails being free of any medical condition or predisposition and being medically ready (where readiness is the capability of being able to accomplish a task)" (Shih, Meadows, and Martin, 2013, p. 5). Medical fitness acts as buffer to stressors, inherently improving well-being: "Medical fitness . . . can be thought of as a tool with which to buffer stress, and lack of medical fitness can be an aggravating factor that makes it harder to cope with stress" (Shih, Meadows, and Martin, 2013, p. 5). Those who are not medically fit might be more likely to respond negatively to stress or unable to cope with stress caused by, for example, physical impediments or a chronic medical condition (Shih, Meadows, and Martin, 2013, p. 5). There are several research-based recommended COAs presented by Shih, Meadows, and Martin, 2013, which we discuss later in this section.

Table C.2 summarizes the soldier well-being components discussed in this section.

Stressors and Detriments to Well-Being

Soldiers stationed at temporary bases abroad experience a range of unique situational stressors in addition to those stressors that military personnel face in everyday military life. Marlowe, 2001, notes, "The stresses of combat, deployment, or high threat environments can, for some military personnel, have both immediate and long-term disruptive physical and psychological consequences" (Marlowe, 2001, p. 1). Understanding these stressors and how they relate to applicable constructs and components of well-being is critical to applying research-based recommendations properly and guiding future policy. Using previous research, we describe under "Gulf War Stressors" an illustrative example documenting many of these unique stressors.

Gulf War Stressors

The 2001 RAND National Defense Research Institute publication *Psychological and Psychosocial Consequences of Combat and Deployment with Special Emphasis on the Gulf War* (Marlowe, 2001) chronicles extensive research on stressors affecting soldiers deployed during operations Desert Shield and Desert Storm. The RAND study focused on combat arms units rather than on support groups because the researchers deemed the former to be at higher risk for psychological and psychiatric trauma than the latter.

Several notable infrastructure- and capital-based stressors adversely affected troops. Most of these stressors are unique to temporary bases. The nature of the deployment, in which soldiers were isolated from the host nation almost entirely and temporary bases were constructed away from cities and the amenities that accompany them, left soldiers without chances to interact with host cultures or the population:

> Drawing from soldiers' statements in our interviews in the Gulf, we find that many of the usual mediating structures that buffered previous deployments were absent, particularly those provided by interaction with the host society and the opportunities for ease, recreation, diversion, and relief from military routine that such interactions provide. (Marlowe, 2001, p. 116)

Furthermore, bases seemed to be hastily built, overcrowded, and without much room for privacy or personal time:

> [In Saudi facilities, the] gleaming concrete contemporary buildings and billets were deceptive since there was extensive internal crowding. These facilities were often surrounded by large and equally crowded "tent cities." Despite the "modernity" of these facilities, soldiers were physically and psychologically constricted. For the overwhelming majority of the force, there was no place to go—no place for people to physically escape from each other. The crowding—often considered a major source of stress—and the omnipresence of leaders

gave soldiers and leaders both a sense of constant unbroken evaluation of each by the other to the discomfort of both. (Marlowe, 2001, p. 125)

Marlowe stresses the proximity of leaders to soldiers, finding that constant contact and cohabitation of leaders and soldiers left both leaders and soldiers on edge, feeling unable to be themselves or "let their hair down." Marlowe, 2001, heavily stresses issues with living space:

Living space was, overall, a chronic and persistent stressor.[3] . . . At this time, the common denominator of life for most of our troops was crowding amid vast emptiness and fairly complete isolation in proximity to modern cities. In a land of limitless horizons, many of our military personnel lived lives as constricted as sardines in a can. (pp. 126–127)

A lack of MWR equipment compounded the effects of isolation. Not only could troops not get away from each other; they could not use their time to relax with each other as well:

There was little MWR (morale, welfare, recreation) equipment available. If the soldiers' military organization had not brought things with them into the theater on the initial deployment, there were few books, athletic equipment, movies, etc. available. They were, at best, in transit. . . . Life was almost completely restricted to what could be done within the very limited areas inside the wire, fence, or walls that encircled the base. (Marlowe, 2001, p. 127)

Issues with living space, a lack of interaction with local culture, the general location of theater (the Arabian Desert), and an additional lack of capital to help offset these issues on base (MWR equipment) provided high levels of stress and drove a sense of "cabin fever" among troops.

Other points of infrastructure deficiencies contributed to reduced morale and presented additional stressors. A lack of reliable mail service and telephone communication infrastructure contributed heavily to unease and increased uncertainty surrounding soldiers' personal lives back home:

Mail service was appallingly slow. Communication to and from home was measured in weeks, rather than days for the regular mail or seconds for telephone service that Americans have come to expect. For Gulf-deployed Army personnel, telephone communication was difficult to impossible. For most it was simply not available. Therefore, problems at home could not be understood and dealt with in reasonable time frames. In some cases, this was underlined by beliefs, both true and fantasized, about potential dangers both to family members and property at home. For a number in the Army, the salience of these issues was increased by the perception of a status of relative deprivation compared to other services— it was common knowledge that Air Force personnel had regular phone contact with home while soldiers did not. (Marlowe, 2001, pp. 123–124)

Marlowe additionally makes note that Army personnel felt relatively deprived, given that Air Force personnel had these infrastructure elements in place.

---

[3] Footnote in the original: "This was particularly true for those in the combat arms. Many of these soldiers were both socially and physically crowded together while living in the midst of an empty desert—an environment of almost utter barrenness."

Not all stressors found were infrastructure related. Many were psychological, exacerbated by a lack of accurate information, rumors, tensions and suspicions, and regional restrictions. The psychological effects of the Gulf War tended to be unlike those seen in previous U.S. combat engagements—although a relatively small proportion of soldiers required medical attention, unique psychological conditions appeared within a subset of the deployed population following operations Desert Shield and Desert Storm (Marlowe, 2001, pp. 115–116). Marlowe characterized the general nature of soldiers with whom he had met who deployed to the Middle East during the Gulf War as follows:

> [M]orale was good to high in most cases, and soldiers performed effectively and with humor. But unease and anxiety lurked continuously at the edges, frustration levels were apparently high, and the soldiers' universe was persistently seen and felt as threatening. (p. 116)

> There was a constant undertone of apprehension and stress, ranging from vague unease to strong overt apprehension, referable to an ambiguous environment—superficially peaceful, but with constant lurking threats. (p. 122)

It is the sense of underlying, continuing, and chronic stress that lies at the heart of the psychological and psychiatric consequences of Gulf War deployment of soldiers whom Marlowe studied. Psychological doubts and uncertainty about U.S. capabilities in the face of what was perceived as a formidable enemy contributed to high levels of stress among combat units:

> Prior to the Gulf War, public doubt about our weapons systems was a matter of common discussion. . . . American forces were also untried; it had been a full generation since Americans had been involved in a major conflict. . . . Complementing the vision of a vulnerable and "victimizable" American force was that of an almost invincible Iraqi military. . . . It was said to be dedicated, cohesive, and well led, and it was assumed to possess high-technological capacity acquired from the former Soviet Union. . . . Major doubts were circulated in the media and in both military and civilian rumor systems about the quality and effectiveness of the U.S. military's gas warfare protective gear and gas detection systems. . . . Such perceptions of potential defects in our equipment and of enemy strengths were present at the beginning of the deployment, and our interviews showed that they remained part of the soldiers' mental baggage throughout Desert Shield and Desert Storm. These concerns were a continuous aspect of the context of the lives of the deployed troops, reinforced by radio newscasts, CNN, and English-language newspapers and magazines. (pp. 117–118)

The rumor and media mill surrounding operations in the Middle East led to increased stress among combat units, fostering insecurities and growing feelings of apprehension. Perceptions of enemy superiority contributed as a stressor among soldiers stationed abroad. Contributing to these perceptions were indications, through letters from and contact with home, that the United States was expected to suffer "extraordinarily high casualties" if war came and that "the force was badly outnumbered and outgunned" (Marlowe, 2001, p. 128). Furthermore, "[i]n a number of venues, soldiers went on their physical training runs in helmets and Kevlar body armor. In some, they were required to wear body armor throughout the day" (p. 128).

These requirements contributed to general discomfort and stress from a physical standpoint but also weighed as psychological stressors, promoting a sense of imminent or impending conflict.

Marlowe found that concerns about deployment were prevalent among responses to surveys: "The most salient stressor during this initial period was the high level of concern about a projected date of return to the United States" (p. 124). Uncertainty surrounding return dates from deployment, overall military strategy, and deployment length acted as significant stressors (p. 123). Deploying the first majority-married professional Army (more than 60 percent of the force deployed to the Gulf was married) increased stress levels pertaining to deployment uncertainty because married soldiers were eager to return to spouses, children, and family (p. 124).

Additional stressors pertained to the deployment location. Climate and desert and austere conditions were noted as significant stressors, given extreme temperatures, sand and winds, flies, and other factors, although these were noted with less frequency than stressors discussed previously. Some stressors were related to cultural compromises that had to be made to allow for the operation of temporary bases in host nations. An inability to practice religion was an issue for some soldiers. Issues with women having to wear long-sleeved shirts during the desert heat while men could wear cooler, short-sleeved (or no) shirts, became an issue for women stationed at temporary bases in the Middle East: "Many women, in particular, and some men, described a sense of identity loss at not being able to dress and act like Americans" (Marlowe, 2001, p. 127).

Organizational requirements also acted as stressors: Long days of work (14 or 15 hours per day for six or seven days a week in many cases), sleep deprivation, and severe physical stress contributed to the overall sense of anxiety. The use of "fillers," or personnel who were rapidly deployed to fill voids in units and lacked cohesion and rapport with current unit members, contributed to a sense of confusion or lack of cohesion, increasing stress. Troops in some organizations were still eating MREs—Army emergency combat rations—months after arriving in the field (Marlowe, 2001, pp. 129–130).

During combat, the use of pyridostigmine bromide was a large stressor: Soldiers claimed that they were "guinea pigs" taking "untried, experimental drugs" (Marlowe, 2001, p. 140). These perceptions led to fears of unintended side effects (or that side effects were not well understood) and potential consequences, increasing stress levels. Many did not wish to be guinea pigs and recalled Agent Orange in many instances (p. 141). Interestingly enough, however, combat served a dichotomous purpose for soldiers engaging:

> Combat was perceived as both a stress reliever and as a source of stress. In interview, soldiers commonly noted that crossing the [line of departure] was the greatest stress reliever of the entire deployment, since it meant they were on their way home. Combat was of course also a source of apprehension; counterbalancing the apprehension was a strong sense of obligation. (p. 141)

However, normal combat stressors were reported as well. Having a buddy killed or wounded in action, being wounded or injured, having a leader killed or wounded in action, having a confirmed enemy kill, seeing an enemy soldier killed or wounded, and being attacked by enemy artillery or tanks were all significant combat-related stressors that Marlowe reported. Additionally, anticipation of combat led to anxiety, though it was mitigated by the notion that combat would mean returning home soon (p. 134) and "obligations to each other and the mission came first" and by some coping strategies, such as smoking cigarettes (p. 136).

Some mitigation and stress-remediation tactics are fairly intuitive—provide less crowded living arrangements; ensure that living arrangements are constructed for soldiers before they arrive; ensure that MWR equipment is available—though some are not—dealing with the desert and providing interaction with local populations and culture when they might be located far from population centers, for example. Related to stress-mediation recommendations and methods are the lack of coping mechanisms that would be available to soldiers in civilian life. For example, the inability to consume alcohol acted as a stressor, not for physiological reasons but because of the inability to engage in off-duty relaxation or "unwind after work."

Reserve and National Guard units expressed stress over treatment at temporary forts abroad. Many pointed out problems with equipment, and some complained about how they were treated, noting that they were feeling as though they were not well integrated into facilities and that messes were not being shared appropriately; furthermore, because National Guard units did not have the same family support benefits that active-duty military members do, there was heightened stress regarding families, careers, and finances (Marlowe, 2001, pp. 135–136).

In three tables in his report, Marlowe includes full lists of stressors found; however, he does not address every stressor in his commentary. Therefore, we include the following two lists—which denote responses found in Marlowe's Tables 10.2 through 10.4—of responses and how they fit into Marlowe's primary themes and do not list them in our summary tables at the end of this appendix. The first list of these stressors is provided as follows, from highest percentage of respondents to lowest:[4]

> not having the opposite sex around; flies; lack of contact with your family back home; not having private time; not being able to act like Americans; eating MREs a lot of the time; people in other units having things better than you; having your leaders around too much; lack of alcoholic beverages; lack of adequate MWR; illness or problems in your family back home; lack of understanding about why you were deployed to the Middle East; eating T-Rations [tray rations, one kind of B-ration] a lot of the time; unusually long duty days; having to do extra details; not being able to stay in shape; being at MOPP [mission-oriented protective posture] level 3 [wearing suits, boots, and masks and carrying gloves] or 4 [wearing all protection] for a long period of time; maintaining equipment in desert operations; behavior restrictions in the presence of the Saudis; lack of confidence in MOPP gear; what you see or hear on TV or radio about Operation Desert Storm; operating in desert heat; crowding at base camps; fights or quarrels among soldiers in your squad/section or platoon; not getting enough sleep; talk about projected cuts in Army strength; talk about reductions in force in my pay grade; what your family members write to you about Operation Desert Storm; operating in the desert sand; not being able to accomplish your mission while wearing MOPP gear; scorpions, snakes, and spiders; severe change in temperature from day to night; terrorist attacks; not having enough physical energy to do your job; talk about [quality military performance]; shift work; not having bottled water; desert storms; having to train at night; not having time or place to practice your religion; terrorist threat; not being allowed to practice your religion because of host nation restrictions; becoming dehydrated. (Marlowe, 2001, pp. 133–134)

Secondly, a significant number of responses acted as both stressors and coping mechanisms. Because Marlowe, 2001, does not discuss many of them in significant depth in his

---

[4] From Marlowe, 2001, Tables 10.2 and 10.3 (pp. 133 and 134, respectively), "Questions to Which 16 Percent or Less of the Sample Responded This Caused 'Quite a Bit' or 'Extreme' Stress During the Past Week" and "Questions to Which over 16 Percent of the Sample Responded This Caused 'Quite a Bit' or 'Extreme' Stress During the Past Week."

description of findings, we provide these responses only as a list;[5] again, we have organized response categories from the highest percentage to lowest percentage of respondents, noting that the categories caused stress (which is *generally* inversely related to the percentage for which the same category acts as a coping mechanism):

> length of tour; not knowing if we will go into combat; what I think the Iraqis might do; sanitary conditions; chain of command; length of time between field rotations; present living conditions; lack of variety in things to do; rumors; command information; family problems; training; other soldiers in platoon/squad; heat and climate; information about Iraq; choices as to how I spend my time; health concerns; improvements in living conditions; newspapers; letters from home; phone calls home; rest days; trips to rest areas; Armed Forces Network radio; entertainment we create; cold sodas/munchies; chaplain visits; sports; watching TV; reading books. (Marlowe, 2001, p. 135)

For example, 69.6 percent of respondents noted that the length of tour caused stress, while 8.4 percent who stated that it helped them cope; 6.2 percent of respondents noted that reading books caused stress, versus 55.0 percent who stated that it helped them cope.

Additional Stressors

Additional stressors are described throughout the literature reviewed. Robson, 2013, notes that, although physical activity and exercise are largely beneficial to anyone engaging in them, drawbacks can apply to military lifestyles. As Robson states, "there are some identifiable risks associated with exercise. Disruption in regular patterns of exercise can elicit a range of negative emotions" (p. 14).[6] Additionally, "exercise at intensities beyond usual levels may deteriorate mood," while exercise can lead to an increase in injuries (p. 14). Injuries can then act as a stressor, as noted by Shih, Meadows, and Martin, 2013.

Medical Detriments to Well-Being

The detrimental effects of some medical conditions, ailments, and injuries on well-being are well known. Shih, Meadows, and Martin, 2013, presents three applicable constructs discussing medical detriments to well-being: facilitators and barriers to health care, chronic health conditions, and the presence and management of injuries.

Facilitators and Barriers to Health Care
Facilitators and barriers to health care are not of significant relevance to soldiers stationed at temporary bases abroad, but there are applications. Shih, Meadows, and Martin, 2013, describes research from several sources that indicates issues with barriers to care—especially

---

[5] From Marlowe, 2001, Table 10.4 (p. 135), "Categories Perceived as Sources of Stress of as Coping Aids by XVIIIth Airborne Sample."

[6] It should be noted that there are concerns over some studies that examine the effects of exercise withdrawal; notably, there might be a difference in the population of those who would volunteer for an exercise-withdrawal study and the general population of exercisers (Robson, 2013; Robson, 2014).

mental health care—affecting soldiers in theater: "A number of recent studies have reported that perceived stigma is a barrier to receiving psychological care among veterans and soldiers in theater" (p. 19). Of note is that

> service members who screen positive for psychological problems are likely to have stronger levels of stigma [and] perceptions of stigma with seeking care, such as feeling embarrassed or worrying that it would harm one's career or be seen as weak, were reported twice as frequently among soldiers in theater who screened positive for mental health problems as among those who did not. (Shih, Meadows, and Martin, 2013, p. 20)

Chronic Health Conditions

Chronic health conditions can hinder well-being and "may limit an individual's ability to be resilient in the face of stress" (Marlowe, 2001, p. 28), while "[s]ome chronic health conditions can interfere with job performance, readiness, and quality of life" (p. 21). Chronic conditions of growing concern affecting U.S. Air Force airmen—and potentially the Army—today include overweight and obesity, diabetes, and asthma (p. 21). Research indicates that obesity is also associated with many qualities detrimental to resilience and fitness, which are components of well-being, in general:

> Obesity is associated with the development of depressive symptoms, self-stigma, reduced quality of life, and severe isolation. . . . In addition, research has shown that individuals who are unable to maintain weight loss are more likely to have a narrow range of coping skills. . . . For example, when obese individuals are exposed to stress or negative emotions, they tend to employ avoidant or impulsive styles of coping, such as escape/avoidance, eating to regulate mood or distract, smoking, taking drugs, taking tranquilizers, sleeping more, and wishing problems would resolve themselves. (Marlowe, 2001, p. 24)

Obesity is associated with numerous and chronic health conditions and is of growing concern for branches of the military given increases in rates of obesity among those eligible for military service; furthermore, it is the leading cause of failing applicants who otherwise qualify for military service (Marlowe, 2001, p. 23). Obesity is associated with mental health conditions, "although the direction of causality has been difficult to establish" (p. 23).

Diabetes is detrimental to well-being through general detriment to health and an associated loss of productivity—Shih, Meadows, and Martin, 2013, cited research (from Stewart et al., 2007) that found

> health-related lost productive time was 18 percent higher in diabetic adults who reported neuropathic symptoms, such as tingling hands or feet and numbness. . . . The study estimated that those who worked with diabetic neuropathic symptoms lost 1.4 hours of work per week . . . . (p. 26)

Finally, asthma is a chronic condition of concern to the U.S. military. According to Shih, Meadows, and Martin, 2013,

> Research indicates that asthma can be affected by anxiety, stress, sadness, environmental irritants, or allergens, exercise, and infection [and] is often correlated with anxiety and

depressive disorders. . . . Uncontrolled asthma can lead to lower quality of life, future lung damage, and even mortality. (p. 27)

Injuries

Injuries are another stressor that Shih, Meadows, and Martin, 2013, addresses. Sommers, 2006, defines injury as "the physical damage resulting from exposure of the human body to sudden intolerable levels of energy" (Shih, Meadows, and Martin, 2013, p. 29). Clearly, injuries are detrimental to overall well-being and QoL because they can range from minor and incidental to life-threatening and fatal. Research has shown that traumatic injury outcomes include

> long-term loss of productivity in both society and the workplace, a high incidence of psychological symptoms, [and] a link between poor recovery and increased drug and alcohol consumption [as well as associations with] depression, PTSD [posttraumatic stress disorder], lower rates of returning to work, lower general health, lower quality of life, and lower overall satisfaction with recovery. (p. 30)

Additionally, of relevant concern is the fact that research considering multiple combat injuries, or polytrauma, has indicated that (in comparing patients with and without polytrauma) physical functioning in those with polytrauma was significantly more impaired 12 months posttrauma (p. 31).

Traumatic brain injury is clearly detrimental to well-being because it can affect patients in any one or more of these three distinct ways: physical, cognitive, and behavioral or emotional (Shih, Meadows, and Martin, 2013, p. 32). It has been linked to numerous outcomes and conditions that are detrimental to soldier well-being, including "depression and anxiety, low return to productivity, loss of independence, reductions in social networks, personality and behavioral changes, chronic pain, suicide ideation and completion, and use of avoidant coping" (p. 32). Similar associations are found in patients with chronic pain.

Recommendations and Means to Improve Well-Being

Given the scope of this literature review, it is important to consider recommendations that researchers propose for remediating the effects of various stressors and conditions that can be detrimental to well-being, especially those that Marlowe, 2001, discusses.

Fitness and Resilience Improvements

The vast majority of recommendations presented in literature focus on improving fitness and resilience and can generally be associated with previously described constructs in each of the three classes examined herein: psychological, physical, and medical.

Psychological Fitness and Resilience

Recommendations and methods to improving well-being with respect to psychological well-being come from both Robson, 2014, and Bates et al., 2010. First, Robson presents few or no recommendations and instead focuses on positive and negative outcomes of different dimensions of each construct with the intention of developing metrics for psychological fitness. These constructs, however, can be seen as useful for understanding which psychological resources are best suited for soldiers stationed at temporary bases given the unique challenges they might face.

Self-Regulation

Robson does not present any recommendations for improving self-regulation and instead focuses on metrics and measurement.

Coping Strategies

Robson presents several coping strategies and associated research examining each strategy. In Israeli soldiers, emotion-focused coping has been found to be a predictor of stress during survival training and of "distress and poor performance in soldiers performing an evacuation task"; those relying on this strategy were significantly more at risk than those using problem-focused coping of developing PTSD symptoms (Robson, pp. 12–13). However, there have been some issues with measuring coping, and emotion-focused coping has been associated with lower levels of PTSD at very high levels of combat exposure (p. 13). Cognitive reappraisal, or changing one's view of a situation to be more positive, is a coping strategy that has been shown to be valuable for coping with stress (p. 13). Additionally, flexibility in coping—i.e., the ability to adapt the coping mechanism to a given situation—has been shown to be valuable in dealing with stress (pp. 13–14).

Positive and Negative Affect

Negative affect has an adverse relationship with well-being, while Lyubomirsky, King, and Diener, 2005, suggests that positive affect has been suggested to be related to "confidence, optimism, and self-efficacy; likability and positive construal of others; sociability, activity, and energy; prosocial behavior; immunity and physical well-being; effective coping with challenge and stress; and originality and flexibility" (Robson, 2014, p. 15).

Perceived Control

Perceived control, also known as locus of control, can be beneficial to well-being. As Robson writes,

> those with an internal [locus of control] might believe that their performance evaluation scores were due to their own efforts. In contrast, individuals with an external [locus of control] might believe that their performance evaluation scores were due to luck or to their assignment to a particular unit of supervisor, resulting in low expectations that increased efforts will lead to higher performance evaluation scores. (p. 16)

Internal locus of control has been associated with "positive task and social experiences and higher levels of motivation, satisfaction . . . performance . . . less intense PTSD symptoms," less stress and higher engagement in task-focused coping mechanisms following natural disasters, and control in one's life and behavior, which is considered a core component of well-being and is associated with "a range of positive psychological states" (Robson, p. 16). External locus of control has been associated with higher risks for depression and anxiety; poor response to stress; less demonstrable happiness; and, in combination with stress, which might lead to a loss of perceived control, higher risks for substance abuse (p. 16).

Self-Efficacy

Self-efficacy has been "positively associated with job performance and satisfaction . . . academic performance and persistence," less stress, "autonomic arousal when attempting to solve challenging problems," and adapting or adjusting to change (Robson, p. 17).

Self-Esteem

Self-esteem has been shown to be strongly correlated with overall well-being, to buffer against anxiety, and to support happiness and resilience (Robson, p. 18). However, as Robson notes, research (from Baumeister et al., 2003) has concluded

> that data are not sufficient to justify development of efforts simply to raise self-esteem. Not only is there a lack of evidence for developing such programs, but there may also be certain risks to such enhancements, as certain categories of high self-esteem (e.g., narcissism) can lead to a variety of such negative outcomes as increased bullying, aggressive retaliatory behavior, and prejudice. (Robson, 2014, pp. 18–19)

Optimism

Optimism is significantly correlated with physical health outcomes (although it acts as a better predictor of subjective health outcomes than objective outcomes) and is positively associated with psychological health, well-being, "post-traumatic growth, increased use of approach coping strategies, and reduced use of avoidance coping strategies" (p. 19).

The emerging constructs that Robson presents—adaptability, self-awareness, mindfulness, and emotional intelligence—do not have large bodies of literature examining applicability to well-being and psychological resilience and fitness. However, they are generally associated with positive psychological resilience outcomes.

Next, we discuss the five internal psychological resources that Bates et al., 2010, presents.

Awareness

According to Bates et al., 2010,

> Self-awareness can be developed over time and has been shown to be a significant factor in inferential processes and intelligence. Individuals must also have situation awareness, or knowledge of what is going on around them for accurately interpreting and attending to appropriate cues in the environment. . . . Emotional awareness includes awareness of one's own emotions and the emotions of others. It has been found to impact psychological resilience and coping, as well as performance. Cognitive awareness, or metacognition, is awareness and regulation of one's cognitive functioning and the factors that affect it. Metacognitive strategies can be employed to manage uncertainty in a situation, and research indicates that using metacognitive strategies can enhance adaptability and on-the-job focus. (pp. 28–29)

Bates et al., 2010, is careful to note that attention control can, and potentially cognitive and situational awareness could, be improved over time with individualized training programs (p. 29).

Beliefs and Appraisals

The resource of beliefs and appraisals that Bates et al., 2010, developed includes the constructs of optimism and self-efficacy as Robson defines them and includes associated benefits. Bates et al., 2010, notes that stress inoculation and training are effective means of increasing self-efficacy. Stress inoculation "attempts to immunize an individual from reacting negatively to stress exposure. This process takes place before experiencing the stressful conditions of concern" (p. 29). Bates et al., 2010, notes that "[t]ough realistic training that approximates actual military operations can be a key method for stress inoculation as well as other psychological benefits related to beliefs and appraisals. These potential additional benefits include a sense of psychological preparedness and self-efficacy" (p. 30).

Coping

Bates et al., 2010, p. 30, discusses several coping strategies. The strategies include

- *problem-focused coping*, which "refers to active efforts to confront and manage situational demands and to reduce the discrepancy between a current situation and a desirable outcome" and has been linked with greater resilience, increased confidence, and enhanced performance
- maladaptive coping, which promotes reducing stress in the short term while potentially increasing long-term stress (some practices include "uncontrolled anger, alcohol abuse, aggression towards others, and self-harm"; these practices have been shown to be counterproductive to increasing resilience and well-being)
- emotion-focused coping, which "involves regulating emotions through a broad range of activities such as seeking emotional support, building emotional awareness, working toward acceptance, and positive reappraisal"
- recharging, which includes "practices to restore energy and counterbalance stress that can offset adverse mood and deteriorating performance" (examples include long recovery periods from work and sufficient "down time" between deployments)
- strategically managing energy, which seeks to

    proactively regulate physical and emotional arousal, can promote optimal performance as well as enhance endurance. Two tools shown to be particularly effective in managing energy include relaxation and energization. These techniques utilize imagery, meditation, and muscle relation to produce marked changes in physiological arousal that can be harnessed to quickly and efficiently conserve as well as maximize energy when needed. . . .

- cognitive load-management techniques, which are

    mental strategies (planning, prioritizing, tracking, executing, chunking) used to achieve more efficient task performance or to manage complex or ambiguous information. . . . prolonged attention to tasks that are mentally taxing without sufficient breaks often results in attention lapses or vigilance decrements.

Decisionmaking

Bates et al., 2010, presents several key factors in decisionmaking and makes recommendations on ways to improve it, which has important resilience and performance implications. Decisionmaking has several key points of operational relevance, including

    increased operational intensity, tempo, and scope, the interrelationships between humans, agencies, and technology, and the uncertainty that places increased value on the human capabilities of quick and accurate thinking, planning, acting, assessing feedback, and modifying plans. The decision making factors include problem solving, goal setting, adaptive thinking, and intuitive thinking. (p. 30)

Decisionmaking factors are all associated with various resilience and performance outcomes, including "effective coping responses to intrapersonal and interpersonal stressors, stress experiences, anxiety, depressive experiences, and more severe clinical forms of distress such as suicidal ideation and hopelessness" (Bates et al., 2010, p. 31). According to Bates et al., 2010,

> Studies have . . . demonstrated that decision making skills are modifiable through training. Training programs that incorporate self-regulatory skills such as metacognition offer a method to enhance adaptive thinking. . . . Incorporating these empirically supported training characteristics in combat training simulators that replicate the extremes of combat in a secure environment . . . offers a method for enhancing desired decision making skills and improving resilience in combat scenarios. (p. 31)

Engagement

Engagement has important, positive consequences for improving responses to stress and performance. There are

> two methods for preserving and increasing engagement. . . . [E]ngagement can be fostered by focusing on a person's strengths [from a management perspective.] A second method is ensuring a balance among resources such as job control, supervisor support, access to information, performance feedback, and social support. (p. 32)

Each of these psychological constructs has apparent value in buffering against and mediating stressors.

Physical Fitness and Resilience

Physical fitness and resilience have been shown to have extensive benefits in mitigating stress, improving overall physical and mental health, and, subsequently, overall well-being. However, as Robson, 2013, notes, "science is moving away from fitness standards based on population norms (e.g., percentiles) to those based on health-related outcomes" (p. ix). Robson summarizes findings on interventions to improve physical fitness as follows:

> Interventions to promote physical fitness are clustered in three areas: informational approaches, behavioral and social approaches, and environmental and policy approaches. Informational approaches are designed to motivate, promote and maintain behavior primarily by targeting cognition and knowledge about physical activity and its benefits. Behavioral and social approaches are designed to foster the development of behavioral management skills and modify the social environment to support changes in behavior. And environmental and policy approaches aim to increase opportunities to be physically active within the community. Ultimately, any policy or program aimed at increasing physical activity should recognize that fitness habits are the result of demographic (e.g., gender, age, ethnicity), psychological, lifestyle, and environmental factors. The decision to exercise and to maintain an exercise program often depends on a number of these factors, so an intervention with a single focus may not be as effective as a multifaceted approach. (p. ix)

The benefits of physical activity are numerous and well-known to the Army. Sedentary lifestyles are more likely to experience "chronic health conditions, including breast cancer, depression, hypertension, lower quality of life . . . coronary artery disease [and] large economic costs . . . ." (Robson, 2013, p. 12).

In addition to the benefits of physical fitness, activity, and resilience outlined previously, mental health benefits of physical activity and fitness are well known. A strong body of evidence indicates that, in responsive and sustained scenarios, regular physical exercise has a mediating effect on stress. According to Robson,

> exercise enhances neuronal functioning by facilitating the expression of certain neurotrophic and neurogenic factors, such as brain-derived neurotrophic factor. These factors support developments of the nervous system, maintenance of neurons in the brain, and neural plasticity [from Duman, 2005] [and] can protect against the onset of certain mental health disorders and symptoms, including depression and anxiety. (Robson, 2013, p. 14)

Recommendations include utilizing physical fitness, though scientific and civilian literature points to health outcomes over fitness (Robson, 2013, p. ix). Physical fitness is more amenable to objective assessment, though issues include factors affecting fitness exams (e.g., larger people have a more difficult time doing pull-ups than smaller people do, "even if they are not fatter") and that genetic inheritance affects cardiorespiratory fitness exams (p. 16). Recommendations include careful consideration when selecting a measurement approach.

Physical fitness is important to the mission of the military and, for a variety of reasons, has strong effect on the well-being of troops stationed at temporary bases, whether it pertains to job performance, the demands and rigors of deployment, or general well-being. Indeed, Marlowe, 2001, found that 23.2 percent of respondents polled as part of his study of the psychological impacts of the Gulf War responded that "not being able to stay in shape" caused "quite a bit" or "extreme" stress during the past week.[7]

Medical Fitness and Resilience

Shih, Meadows, and Martin, 2013, presents several recommendations from the literature on improving medical resilience, fitness, and consequent well-being. These recommendations fall largely within the scope of the well-being constructs. Preventive screenings, although valuable in effectively reducing many diseases and associated burdens, have no strong body of empirical evidence linking them to airman readiness (and potentially to Army soldier readiness) (p. 15).

Barriers to accessing health care are not as prevalent in the military community as they are in the civilian world because military members receive medical insurance as part of their benefits. However, some factors can contribute to barriers in the military lifestyle. As Shih, Meadows, and Martin, 2013, notes, "Patients who have a usual source of care are . . . more likely to receive preventive services, especially if they also have insurance" (p. 18). Because military members tend to move with fairly high frequency, subsequent consistency issues might be considered a barrier to accessing health care.

Shih, Meadows, and Martin, 2013, describe two additional important considerations for promoting access to care. First, a strong social support network has been shown to increase perceived access to health care and foster a "sense of community in access to health care" (p. 19).

---

[7] Surveys were administered in the Persian Gulf in late 1990 to deployed troops.

Secondly, in recent research, veterans and soldiers have reported perceived stigma as a barrier to accessing health care (p. 19).

Shih, Meadows, and Martin, 2013, summarizes recommendations to improving access to care in a military context as "[a]cting on factors that are more mutable, such as reducing transportation barriers, and introducing higher-level policies and programs, such as evidence-based stigma reduction programs, may be good interventions to addressing barriers to accessing quality medical care" (p. 20).

Recommendations for improving medical fitness with respect to the constructs of chronic conditions and injuries focus mainly on primary and secondary preventive health care, though they incorporate additional care, including holistic or multidisciplinary treatment strategies, cognitive behavioral therapy (pain and disability management), and supplementary care, such as supportive counseling (for patients and families), education, reassurance, and additional short-term psychological care. However, these constructs tend to be issues that the military as a whole faces and are not unique to temporary bases.

Illustrative Example: Gulf War

With respect to the illustrative example of unique stressors to soldiers deployed during the Gulf War, several mediation steps relieved stress, though many of them were fairly intuitive and non-novel. Marlowe, 2001, notes,

> As the deployment progressed, a number of the initial sources of stress were dealt with and moderated as the theater infrastructure matured. While crowding continued, the supply of tentage and cots was adequate for all troops by the November–December time period. T-packs, supplemented by at least some Class A rations, were not widely available, and, in the field, MREs were seldom utilized for more than one meal a day. By the end of November, telephone banks had solved a number of the problems of communication with home but had by no means resolved all the worry and strain. Unit recreation centers with large screen television sets and athletic equipment offered at least partial escape from the omnipresent chain of command in most organizations. The cruise ships, docked at Bahrain, the rest and recreation facility at Half Moon Bay, and ARAMCO "home visit" program all offered relief to part of the force. A Thanksgiving dinner, including a presidential visit and celebration, provided another moment of relief.

> One important resolution was the end of the fear of being treated with the same rejection experienced by Vietnam veterans. The outpouring of support from the American people was truly meaningful to the troops. Soldiers were deluged by "any soldier" letters, boxes of cookies and candy, and a tidal wave of other messages of support. The massive arrival of new equipment, the imminent arrival of VIIth corps, and the military edge represented by the M1A1 Abrams tank all contributed to an easing about the capability of dealing with Iraq's heavy armored forces. (Marlowe, 2001, p. 131)

As Marlowe, 2001, notes, much of the stress that initially struck soldiers was mitigated as infrastructure grew over time. As discussed previously, combat was both a stressor and a stress reliever because engaging in combat meant that the soldier was soon to return home. In general, much of the stress appears to be related to a lack of suitable infrastructure and amenities (having facilities in place with sufficient capacity can partially mitigate privacy and crowding concerns), a lack of normalcy and stress mediators that are available at home, family and support systems, combat, and uncertainty regarding deployment times.

Additional Recommendations

Additional recommendations focus on improving understanding of well-being and QoL as a whole. We find these recommendations primarily in Sims et al., 2013. Sims et al., 2013, finds that research needs to be improved and expanded, especially in understanding QoL in general and how it relates to the Army's strategic goals. The primary conclusions of the report are as follows (pp. xiv–xv):

- "some prerequisites for a rigorous roadmapping exercise (to determine the course of future QoL research) are missing"
- "domain-specific research remains central to developing solutions and assessing their effectiveness"
- "assessing QoL needs a big-picture understanding of stressors within and across multiple domains in life"
- "current needs assessments are not broad enough."

Furthermore, the authors present tangible recommendations to improve the understanding of QoL and how it pertains to the Army's overall strategic goals going forward (pp. xv–xvi):

- "develop an agreed-upon QoL lexicon, outcomes, and metrics"
- "focus research on individual domains to build the big picture"
- "take a comprehensive approach to needs assessment"
- "improve knowledge management to expand research use and [identify] important areas for new research"
- "make Army QoL research roadmapping a socialization and knowledge-sharing process"
- "target research in areas where the army can make a difference."

Performing research into improving QoL for soldiers stationed at temporary bases abroad while keeping these recommendations in mind can help to advance the Army's overall QoL goals and intended strategic outcomes. Maintaining a unified QoL lexicon is important to understanding the connection between QoL and well-being for these soldiers. Working from this lexicon, improving knowledge management, and engaging in knowledge sharing are all potential areas in which this research fits within these constructs.

Summary Tables

In this section, we provide summary tables of information regarding stressors and mitigation recommendations based on applicable constructs. Table C.3 summarizes psychological fitness and resilience constructs, along with applicable information on the benefits of increasing these constructs and recommendations, if applicable;[8] Table C.4 shows stressors as described in Marlowe, 2001; Bates et al., 2010; Robson, 2013; and Shih, Meadows, and Martin, 2013; along with applicable mediators and recommendations.

---

[8] Measurements and metric information for each psychological construct might be available in the sources listed (not all constructs have well-defined metrics).

Table C.3
Summary of Psychological Resilience and Fitness Constructs, Including Benefits, Drawbacks, and Enhancement Recommendations and Implementations

Construct	Benefit	Drawback	Enhancement Recommendation or Implementation
Self-regulation[a]	Ability to exercise restraint, make direct choices, face adversity, and recover from stress		
Coping strategies[a]	Enhanced response to stress		Flexibility in coping (e.g., switching between emotion-focused coping and cognitive reappraisal)
Positive and negative affect[a]	Suggested positive relationship with numerous overall well-being and resilience constructs	Negative affect is positively correlated with well-being.	
Perceived control[a]	Positive work outcomes; enhanced response to stress; task-focused coping; additional motivation; emotional well-being; higher self-esteem; better overall health; less-intense PTSD		
Self-efficacy[a]	Enhanced response to stress		
Self-esteem[a]	Positive association with satisfaction and subjective well-being	Insufficient support for self-esteem–raising efforts; narcissism leads to negative outcomes	
Optimism[a]	Significantly correlated with physical health outcomes; positively associated with psychological health and well-being	Maladaptive to coping with persistent stressors	
Adaptability[a]	Aids in stress response; adjusting to new life and work roles and job demands	No clear definition in literature	
Self-awareness[a]	Positive correlation with building resilience, self-esteem, and mental health and protecting against anxiety and depression		
Mindfulness[a]	Potential roles in adaptive decisionmaking, well-being, and reducing mood disturbance and stress (in cancer patients)		
Awareness[b]	Can be developed over time; "significant factor in inferential processes and intelligence [and] psychological resilience and coping" (Bates et al., 2010, pp. 28–29) and performance		Potentially can be improved over time with individualized training programs

Table C.3—Continued

Construct	Benefit	Drawback	Enhancement Recommendation or Implementation
Beliefs and appraisals[b]	See Robson, 2014, on optimism and on self-efficacy.		Stress inoculation (e.g., "realistic training that approximates actual military operations") (Bates et al., 2010, p. 30)
Coping	Response to stress		
Problem-focused[b]	Greater resilience; increased confidence; enhanced performance		"[A]ctive efforts to confront and manage situational demands and to reduce the discrepancy between a current situation and a desirable outcome" (Bates et al., 2010, p. 30)
Maladaptive[b]	Short-term stress reduction	Counterproductive to increasing resilience and well-being; can include "uncontrolled anger, alcohol abuse, aggression toward others, and self-harm" (Bates et al., 2010, p. 30)	
Emotion-focused[b]			Seeking emotional support; building emotional awareness; working toward acceptance; positive reappraisal
Recharging[b]	Restores energy; counterbalances stress		Long recovery periods; sufficient "down-time" between deployments
Strategically managing energy[b]	Promotes optimal performance; enhances endurance		Utilize relaxation and energization, including imagery, meditation, and muscle relations
Cognitive load management[b]	More-efficient task performance; better management of complex or ambiguous information		Taking sufficient, applicable breaks
Decisionmaking[b]	"[E]ffective coping responses to intrapersonal and interpersonal stressors, stress experiences, anxiety, depressive experiences, and more severe clinical forms of distress such as suicidal ideation and hopelessness" (Bates et al., 2010, p. 31)		Training programs (see description in Section IV)
Engagement[b]	Improved response to stress performance		Focusing on a person's strengths; ensuring a balance among resources for an individual

[a] Robson, 2014.

[b] Bates et al., 2010.

Table C.4
Summary of Well-Being Detriments and Stressors for Gulf War Case Study and Medical and Physical Fitness and Resilience Stressors

Stressor	Mediator	Recommendation
Gulf War case study		
Infrastructure deficiencies	Building infrastructure	
Facilities, living space, and overcrowding[a]	Infrastructure; mitigation by obtaining adequate supply of tents and cots	
No or inadequate MWR opportunities[a]	Building recreation rooms with TVs and athletic equipment; docked cruise ships with MWR amenities and R&R facilities	
Mail and communication infrastructure deficiencies[a]	Increasing available telephone banks and infrastructure	
Worrying and stress from inability to communicate[a]	Increasing available telephone banks and infrastructure	
Host-nation issues		
No or inadequate local interaction[a]		
Deployment location (desert, extreme climate, dangers posed by local fauna)[a]		
Loss of identity by not being able to "act like an American"[a]		
Rumors, tensions, and suspicions		
Doubts about force capabilities[a]	Arrival of new equipment; military edge represented by tanks; arrival of airborne	
Perceived enemy strength[a]	Arrival of new equipment; military edge represented by tanks; arrival of airborne	
Media[a]		
Combat anticipation[a]	Smoking cigarettes; obligations to team and mission	
Other stressors		
Uncertainty surrounding deployment[a]	Entering combat	
Long workdays[a]		
Use of fillers (rapidly deployed personnel unfamiliar with units)[a]		
Extended use of MREs[a]	Obtaining class A rations	
Combat stress[a]		

Table C.4—Continued

Stressor	Mediator	Recommendation
(RC only) Not feeling fully integrated into bases[a]		
(RC only) Heightened stress because the RC does not have the same benefits as the active component has[a]		
Other[a]	Thanksgiving dinner, including a presidential visit and celebration; outpouring of support from the United States indicating that soldiers returning would not face Vietnam-era problems	
Adverse conditions surrounding lack of physical fitness and resilience (applicable mental and physical health issues)[b]	Focusing on health rather than fitness	Promoting physical fitness through "informational approaches; behavioral and social approaches; and environmental and policy approaches" (Robson, 2013, p. ix); utilizing a health-based approach instead of a fitness-based approach
Medical fitness and resilience stressors		
Facilitators and barriers to health care[c]	Increased consistency among sources of preventive services; "strong sense of community in access to health care" (Shih, Meadows, and Martin, 2013, p. 19); reduced stigma	"[R]educing transportation barriers [and] introducing higher-level policies and programs, such as evidence-based stigma reduction programs" (Shih, Meadows, and Martin, 2013, p. 20)
Chronic health conditions[c]		Focus on primary and secondary preventive health care, with incorporation of additional care, including holistic or multidisciplinary treatment strategies, cognitive behavioral therapy (pain and disability management), and supplementary care that includes supportive counseling (for patients and families), education, reassurance, and additional short-term psychological care
Injuries[c]		

[a] Marlowe, 2001.
[b] Robson, 2013.
[c] Shih, Meadows, and Martin, 2013.

Shelter Options

There are multiple billeting shelter options in the Army inventory. A sample of tactical shelters is listed in *Guide for Tactical Training Bases, Shelters Handbook* (NSRDEC, 2014). We do not intend to provide an exhaustive comparison of shelter options. Three shelters presented in this appendix, TEMPER tents (see Figure D.1), B-huts (see Figure D.2), and CHUs (see Figure D.3), were commonly used in Afghanistan and Iraq for billeting purposes. SIP huts are similar to B-huts but use insulated panels to dramatically improve their fuel efficiency.

In Iraq and Afghanistan, tentage was used primarily in smaller camps and much less often in medium to large camps. Most of the medium and large camps used B-huts or, more commonly, CHUs.

Depending on the tent size, four to six people can set up a tent; setup or tear-down can take about 30 to 60 minutes. Tents can have single-ply liners or more-insulated liners. They come in a variety of sizes. Sleeping arrangements are bunks that can house up to 12 people in the case of the TEMPER tent, which is used in the Army's force-provider package.

B-huts are semi-permanent plywood structures that hold up to eight people and are expected to last three to four years. The huts can be constructed from raw materials (boards, plywood, and nails) and can take up to almost 1,000 person-hours to build.

CHUs are metal containers that are easily shipped in ocean carriers. They are durable and can last up to 20 years. CHUs can have insulation in floors and ceilings. Soldiers prefer them because of their relative privacy and space for personal items. A CHU can house two people with a latrine located in the middle or one person of higher rank.

Table D.1 compares costs of the three categories of structures.

Figure D.1
TEMPER Tent

Figure D.2
Barrack Huts

NOTE: SIP huts look much like B-huts but have walls constructed from SIPs
instead of wood.
RAND *RR1298-D.2*

Figure D.3
Containerized Housing Units

RAND *RR1298-D.3*

Table D.1
Shelter Cost Comparisons of Select Shelters

Cost	U.S. Air Force Small-Shelter System or Alaska Small-Shelter System (NSN 8340-01-512-0077, 8430-01-512-0071)	Marine Corps Expeditionary Shelter System (HDT Base-X Model 305) (NSN 8340-01-533-1672, 8340-01-533-1668)	TEMPER Airbeam-Supported Tent Type XXXI (Force Provider) (HDT Airbeam Model 2032) (NSN 8340-01-558-4701, 8340-01-559-3852)	SEA Tent (Wood Frame on Wood Piers, Footing Assembly with Plywood Flooring)	CLU or CHU: Nonexpandable ISO (Prefabricated Living Unit) (NSN 5411-01-294-6390, 5411-01-136-9837)	B-Hut	SIP Hut
Capital costs							
Initial cost, in dollars	14,000–25,000	17,184	23,556	11,339	6,500	4,924	16,000
Personnel per structure	15	12–14	15	6	2	8	8
Deployment costs							
Weight or shipping weight	1,200 lb.	573-lb. package	Shelter: 627 lb.; total package: 692 lb.; crated: 1,200 lb.	6,280 lb.	3,900 lb. (nonexpandable); 5,400 lb. (one-side expandable); 6,900 lb. (two-side expandable)	10,753.6 lb. (including all materials and equipment)	10,626 lb. (including all materials and equipment)
Containers	Transported in individual composite container, 4 to a 463-L pallet; 80 cu. ft. packed volume	2 packages: one is 63 x 38 x 28 in.; the other is 63 x 30 x 20 in.	Natick says that crated dimensions are 46 x 47 x 85 in.; HDT says that they are 72 x 40 x 40 in.	Can be procured locally; materials are class IV and are bulky and not stored in great quantities	1 ISO container	5 in one standard 40-ft. container; materials can be procured locally	2 in one 40-ft. high-cube container (only one can fit in a standard container)
C-17 requirements or sea containers							
Personnel	4	4	3; Natick says 4	4	4	8 (plus contractor possibly needed)	8 (from thesis); 3

Table D.1—Continued

Cost	U.S. Air Force Small-Shelter System or Alaska Small-Shelter System (NSN 8340-01-512-0077, 8430-01-512-0071)	Marine Corps Expeditionary Shelter System (HDT Base-X Model 305) (NSN 8340-01-533-1672, 8340-01-533-1668)	TEMPER Airbeam-Supported Tent Type XXXI (Force Provider) (HDT Airbeam Model 2032) (NSN 8340-01-558-4701, 8340-01-559-3852)	SEA Tent (Wood Frame on Wood Piers, Footing Assembly with Plywood Flooring)	CLU or CHU: Nonexpandable ISO (Prefabricated Living Unit) (NSN 5411-01-294-6390, 5411-01-136-9837)	B-Hut	SIP Hut
Days	30 min.	12 min.	15 min.	263.25 labor-hours (from cost estimator; includes construction and electrical installation)	20 min. for one-side expandable; 30 min. for two-side; or 16 labor-hours (from cost estimator; includes construction and electric)	32 person-days; 2 days (128 person-hours from thesis)	<8 hr. (64 person-hours from thesis); 4 hr.
Sustainment costs							
Duration	Minimum usable life expectancy: 10 years with constant use; 20-year storage life				20-year service	3- to 4-year lifespan	
Energy profile	Equipped to accept 120- or 208-volt alternating-current 60-cycle single-		131,400 kWh per year	87,600 kWh per year	35,040 kWh per year	18,820 kWh per year if not insulated	9,150 kWh per year
Fuel requirement, by structure by time, in gallons per year			5,913	3,902.58	1,561.032	1,882	915
Trucks, convoys, containers			0.0008 resupply convoys per year; 1.1826 fuel tankers per year; 0.047304 convoys	0.0008 resupply convoys per year; 0.780516 fuel tankers per year; 0.03122064 convoys	0.0008 resupply convoys per year; 0.3122 fuel tankers per year; 0.01245 convoys		

Table D.1—Continued

Cost	U.S. Air Force Small-Shelter System or Alaska Small-Shelter System (NSN 8340-01-512-0077, 8430-01-512-0071)	Marine Corps Expeditionary Shelter System (HDT Base-X Model 305) (NSN 8340-01-533-1672, 8340-01-533-1668)	TEMPER Airbeam-Supported Tent Type XXXI (Force Provider) (HDT Airbeam Model 2032) (NSN 8340-01-558-4701, 8340-01-559-3852)	SEA Tent (Wood Frame on Wood Piers, Footing Assembly with Plywood Flooring)	CLU or CHU: Nonexpandable ISO (Prefabricated Living Unit) (NSN 5411-01-294-6390, 5411-01-136-9837)	B-Hut	SIP Hut
Personnel to support, in labor-hours per week for maintenance			1	1	0.5		
Casualties			0.00033613 per year from resupply; 0.001988 per year from fuel	0.00033613 per year from resupply; 0.001312 per year from fuel	0.00003613 per year from resupply; 0.0000523 per year from fuel	Battalion-size base camp can expect 0.10 per year	Battalion-size base camp can expect 0.05 per year
Maintenance, parts and labor costs, in dollars			4,371.26	3,395.47	5,608.48		

NOTE: NSN = national stock number. SEA = Southeast Asia. ISO = International Organization for Standardization. kWh = kilowatt-hour.

References

455th Air Expeditionary Wing, Bagram Airfield, Facebook page, undated. As of February 5, 2018:
https://www.facebook.com/BagramAFG/

AAFES—*See* Army and Air Force Exchange Service.

Applegate, Doug, Center of Standardization for Contingency Facilities, Middle East District, Military Construction Requirements, Standardization, and Integration, "Contingency Base Camps," briefing, September 3, 2013. As of January 26, 2016:
https://mrsi.usace.army.mil/cos-old/med/Base%20Camp%20Standards%20Resources/Base%20Camps%20September%203%202013%20Layouts.pdf

Army and Air Force Exchange Service, *Annual Report 2008*, Dallas, Texas, c. 2009. As of April 24, 2015:
http://www.aafes.com/Images/AboutExchange/PublicAffairs/annual08.pdf

Bandura, Albert, "Self-Efficacy Mechanism in Human Agency," *American Psychologist*, Vol. 37, No. 2, February 1982, pp. 122–147.

Bates, Mark J., Stephen Bowles, Jon Hammermeister, Charlene Stokes, Evette Pinder, Monique Moore, Matthew Fritts, Meena Vythilingam, Todd Yosick, Jeffrey Rhodes, Craig Myatt, Richard Westphal, David Fautua, Paul Hammer, and Greg Burbelo, "Psychological Fitness," *Military Medicine*, Vol. 175, No. 8, August 2010, pp. 21–38. As of January 26, 2016:
http://publications.amsus.org/doi/pdf/10.7205/MILMED-D-10-00073

Baumeister, Roy F., Jennifer D. Campbell, Joachim I. Krueger, and Kathleen D. Vohs, "Does High Self-Esteem Cause Better Performance, Interpersonal Success, Happiness, or Healthier Lifestyles?" *Psychological Science in the Public Interest*, Vol. 4, No. 1, May 2003, pp. 1–44.

Brabham, Daren C., *Crowdsourcing*, Cambridge, Mass.: MIT Press, 2013.

Brown, Kirk Warren, and Richard M. Ryan, "The Benefits of Being Present: Mindfulness and Its Role in Psychological Well-Being," *Journal of Personality and Social Psychology*, Vol. 84, No. 4, 2003, pp. 822–848.

Camp Lemonnier Djibouti's Facebook photo stream. As of January 26, 2016:
https://www.facebook.com/CampLemonnier/photos_stream

Carra, Col (ret.) Jeffrey B., and CWO4 (ret.) David Ray, U.S. Marine Corps, "Evolution of Petroleum Support in the U.S. Central Command Area of Responsibility," *Army Sustainment*, Vol. 42, No. 5, September–October 2010. As of February 5, 2018:
http://www.alu.army.mil/alog/issues/sepoct10/petrol_support.html

Carver, Charles S., and Eddie Harmon-Jones, "Anger Is an Approach-Related Affect: Evidence and Implications," *Psychological Bulletin*, Vol. 135, No. 2, 2009, pp. 183–204.

"Casualty Costs of Fuel and Water Resupply Convoys in Afghanistan and Iraq," *army-technology.com*, February 26, 2010. As of October 27, 2014:
http://www.army-technology.com/features/feature77200/

CENTCOM—*See* U.S. Central Command.

Dziegielewski, Benedykt, Jack C. Kiefer, Eva M. Opitz, Gregory A. Porter, Glen L. Lantz, William B. DeOreo, Peter W. Mayer, and John Olaf Nelson, *Commercial and Institutional End Uses of Water*, Denver, Colo.: American Water Works Association Research Foundation, 2000. As of January 27, 2016: http://www.waterrf.org/PublicReportLibrary/RFR90806_2000_241B.pdf

Faust, CPT John, "Bulk Petroleum Challenges in Afghanistan," *Quartermaster Professional Bulletin*, Spring 2007, pp. 23–25.

Federici, Justine, and Jason S. Augustyn, *Modeling Soldier Quality of Life at Austere Contingency Bases*, U.S. Army Natick Soldier Research, Development and Engineering Center Technical Report, draft, 2015.

Flanagan, John C., "A Research Approach to Improving Our Quality of Life," *American Psychologist*, Vol. 33, No. 2, February 1978, pp. 138–147.

Flórez, Karen Rocío, Regina A. Shih, and Margret T. Martin, *Nutritional Fitness and Resilience: A Review of Relevant Constructs, Measures, and Links to Well-Being*, Santa Monica, Calif.: RAND Corporation, RR-105-AF, 2014. As of April 18, 2018: https://www.rand.org/pubs/research_reports/RR105.html

Fontenot, Gregory, E. J. Degen, and David Tohn, *On Point: The United States Army in Operation Iraqi Freedom*, Fort Leavenworth, Kan.: Combat Studies Institute Press, 2004. As of April 24, 2015: http://usacac.army.mil/cac2/cgsc/carl/download/csipubs/onpointi.pdf

Garvin, Chris, and Jim Codling, "Making Grid Connection Happen," *Military Engineer*, undated.

Headquarters, Department of the Army, *General Engineering*, Washington, D.C., Field Manual 3-34.400 (Field Manual 5-104), December 9, 2008. As of March 16, 2018: https://www.globalsecurity.org/military/library/policy/army/fm/3-34-400/fm3-34-400.pdf

———, Office of the Deputy Chief of Staff for Logistics, *Base Camps for Full Spectrum Operations (2015–2024)*, U.S. Army functional area analysis, final report, March 22, 2010a.

———, *Human Resources Support*, Field Manual 1-0, April 2010b. As of October 27, 2014: http://www.ags.army.mil/Files/fm1_0.pdf

———, *Military Morale, Welfare, and Recreation Programs and Nonappropriated Fund Instrumentalities*, Washington, D.C., Army Regulation 215-1, September 24, 2010c. As of October 27, 2014: http://armypubs.army.mil/epubs/pdf/r215_1.pdf

———, *Commander and Staff Officer Guide*, Army Tactics, Techniques, and Procedures 5-0.1, September 2011. As of January 26, 2016: https://fas.org/irp/doddir/army/attp5-0-1.pdf

———, *RAND Arroyo Center*, Washington, D.C., Army Regulation 5-21, May 25, 2012. As of January 26, 2016: http://www.apd.army.mil/pdffiles/r5_21.pdf

———, *Base Camps*, Army Techniques Publication 3-37.10, Marine Corps Reference Publication 3-17.7N, April 26, 2013a. As of March 16, 2018: http://www.marines.mil/Portals/59/MCRP%203-17.7N%20z.pdf

———, Office of the Deputy Chief of Staff, "Contingency Operations Community of Practice," briefing, October 16, 2013b.

———, Office of the Deputy Chief of Staff, "Base Camp Quality of Life Assessment Update," briefing, November 18, 2013c.

———, Office of the Deputy Chief of Staff, "Operational Contract Support," briefing, January 22, 2014.

———, *General Engineering*, Army Techniques Publication 3-34.40, Field Manual 3-34.400, Marine Corps Warfighting Publication 3-17.7, February 25, 2015. As of April 24, 2015: http://armypubs.army.mil/doctrine/DR_pubs/dr_a/pdf/atp3_34x40.pdf

"History of LOGCAP in Afghanistan: Afghanistan LOGCAP III to LCIV to RSM," briefing slides for unknown audience, February 1, 2014.

HQDA—*See* Headquarters, Department of the Army.

IMCOM—*See* U.S. Army Installation Management Command.

Joint Chiefs of Staff, *Joint Operations*, Washington, D.C., Joint Publication 3-0, August 11, 2011. As of January 26, 2016:
http://www.dtic.mil/doctrine/new_pubs/jp3_0.pdf

———, *DOD Dictionary of Military and Associated Terms*, Washington, D.C., Joint Publication 1-02, February 2018. As of March 16, 2018:
http://www.jcs.mil/Portals/36/Documents/Doctrine/pubs/dictionary.pdf?ver=2018-02-21-153603-643

Karbuz, Sohbet, "Military Oil Consumption in Afghanistan and Iraq," *Sohbet Karbuz on U.S. Military Energy Consumption, Geopolitics, and Energy Security*, June 10, 2006. As of February 5, 2018:
http://karbuz.blogspot.com/2006/06/military-oil-consumption-in.html

Karoly, Paul, "Mechanisms of Self-Regulation: A Systems View," *Annual Review of Psychology*, Vol. 44, 1993, pp. 23–52.

Kerce, Elyse W., *Quality of Life: Meaning, Measurement, and Models*, San Diego, Calif.: Navy Personnel Research and Development Center, AD-A250-813, May 1992. As of January 26, 2016:
http://dtic.mil/cgi-bin/GetTRDoc?AD=ADA250813

Keysar, Elizabeth, "Camp Buehring Net Zero IPR CB Community of Practice," National Defense Center for Energy and Environment, April 2, 2014.

Lyubomirsky, Sonja, Laura King, and Ed Diener, "The Benefits of Frequent Positive Affect: Does Happiness Lead to Success?" *Psychological Bulletin*, Vol. 131, No. 6, 2005, pp. 803–855.

Marlowe, David H., *Psychological and Psychosocial Consequences of Combat and Deployment with Special Emphasis on the Gulf War*, Santa Monica, Calif.: RAND Corporation, MR-1018/11-OSD, 2001. As of January 26, 2016:
http://www.rand.org/pubs/monograph_reports/MR1018z11.html

Maslow, Abraham H., "A Theory of Human Motivation," *Psychological Review*, Vol. 50, No. 4, July 1943, pp. 370–396; revised 1970.

Mason, Raymond V., "Transforming Logistics for a New Era," in Association of the U.S. Army, *Green Book*, October 2013, pp. 171–175. As of January 26, 2016:
http://www.army.mil/e2/c/downloads/315015.pdf

McGene, Juliana, *Social Fitness and Resilience: A Review of Relevant Constructs, Measures, and Links to Well-Being*, Santa Monica, Calif.: RAND Corporation, RR-108-AF, 2013. As of April 18, 2018:
https://www.rand.org/pubs/research_reports/RR108.html

Meadows, Sarah O., Laura L. Miller, and Sean Robson, *Airman and Family Resilience: Lessons from the Scientific Literature*, Santa Monica, Calif.: RAND Corporation, RR-106-AF, 2015. As of February 5, 2018:
https://www.rand.org/pubs/research_reports/RR106.html

Moore, James S., *The U.S. Military's Reliance on Bottled Water During Military Operations*, Norfolk, Va.: Joint Forces Staff College, Joint Advanced Warfighting School, master's thesis, June 17, 2011. As of April 24, 2015:
http://www.dtic.mil/dtic/tr/fulltext/u2/a545433.pdf

Mullen, Michael, "On Total Force Fitness in War and Peace," *Military Medicine*, Vol. 175, Supplement, August 2010, pp. 1–2. As of January 26, 2016:
http://hprc-online.org/files/TotalForceFitness.PDF

NSRDEC—*See* U.S. Army Natick Soldier Research, Development and Engineering Center.

Office of the Deputy Chief of Staff of the Army, G-4, Logistics, home page, last updated January 27, 2016. As of January 27, 2016:
http://www.army.mil/info/organization/unitsandcommands/dcs/g-4/

Office of the Under Secretary of Defense for Personnel and Readiness, *Report of the 1st Quadrennial Quality of Life Review*, May 2004. As of January 26, 2016:
http://download.militaryonesource.mil/12038/MOS/Reports/QQLR.pdf

Prati, Gabriele, and Luca Pietrantoni, "Optimism, Social Support, and Coping Strategies as Factors Contributing to Posttraumatic Growth: A Meta-Analysis," *Journal of Loss and Trauma*, Vol. 14, No. 5, 2009, pp. 364–388.

"Quality of Life," *Definitionsfor*, undated. As of February 5, 2018: https://definitionsfor.com/definition/quality+of+life/

Robson, Sean, *Physical Fitness and Resilience: A Review of Relevant Constructs, Measures, and Links to Well-Being*, Santa Monica, Calif.: RAND Corporation, RR-104-AF, 2013. As of January 26, 2016: http://www.rand.org/pubs/research_reports/RR104.html

———, *Psychological Fitness and Resilience: A Review of Relevant Constructs, Measures, and Links to Well-Being*, Santa Monica, Calif.: RAND Corporation, RR-102-AF, 2014. As of January 26, 2016: http://www.rand.org/pubs/research_reports/RR102.html

Robson, Sean, and Nicholas Salcedo, *Behavioral Fitness and Resilience: A Review of Relevant Constructs, Measures, and Links to Well-Being*, Santa Monica, Calif.: RAND Corporation, RR-103-AF, 2014. As of April 18, 2018: https://www.rand.org/pubs/research_reports/RR103.html

Shelter Technology, Engineering, and Fabrication Directorate, U.S. Army Natick Soldier Research, Development and Engineering Center, *Guide for Tactical Training Bases, Shelters Handbook*, Natick, Mass., October 20, 2008. As of January 29, 2016: http://nsrdec.natick.army.mil/media/print/ShelterGuide.pdf

Shih, Regina A., Sarah O. Meadows, John Mendeloff, and Kirby Bowling, *Environmental Fitness and Resilience: A Review of Relevant Constructs, Measures, and Links to Well-Being*, Santa Monica, Calif.: RAND Corporation, RR-101-AF, 2015. As of April 18, 2018: https://www.rand.org/pubs/research_reports/RR101.html

Shih, Regina A., Sarah O. Meadows, and Margret T. Martin, *Medical Fitness and Resilience: A Review of Relevant Constructs, Measures, and Links to Well-Being*, Santa Monica, Calif.: RAND Corporation, RR-107-AF, 2013. As of January 26, 2016: http://www.rand.org/pubs/research_reports/RR107.html

"Shower Water Reuse Systems Employed at Forward Operating Bases in Afghanistan," *Army Sustainment*, Vol. 44, Issue 2, March–April 2012. As of April 24, 2015: http://www.alu.army.mil/alog/issues/MarApril12/headlines.html

Sims, Carra S., Anny Wong, Sarah H. Bana, and John D. Winkler, *Strategically Aligned Family Research: Supporting Soldier and Family Quality of Life Research for Policy Decisionmaking*, Santa Monica, Calif.: RAND Corporation, TR-1256-A, 2013. As of January 26, 2016: http://www.rand.org/pubs/technical_reports/TR1256.html

Skinner, Ellen A., Kathleen Edge, Jeffrey Altman, and Hayley Sherwood, "Searching for the Structure of Coping: A Review and Critique of Category Systems for Classifying Ways of Coping," *Psychological Bulletin*, Vol. 129, No. 2, March 2003, pp. 216–269.

Sommers, M. S., "Injury as a Global Phenomenon of Concern in Nursing Science," *Journal of Nursing Scholarship*, Vol. 36, No. 4, 2006, pp. 314–320.

Stewart, W. F., J. A. Ricci, E. Chee, A. G. Hirsch, and N. A. Brandenburg, "Lost Productive Time and Costs Due to Diabetes and Diabetic Neuropathic Pain in the US Workforce," *Journal of Occupational and Environmental Medicine*, Vol. 49, No. 6, June 2007, pp. 672–679.

TRADOC—*See* U.S. Army Training and Doctrine Command.

USAREUR—*See* U.S. Army Europe.

U.S. Army, *Field and Garrison Furnishings and Equipment*, Common Table of Allowances 50-909, August 1, 1993.

U.S. Army Central, *USARCENT G1 Rest and Recuperation (R&R) Leave Policy and Procedures*, January 2011. As of January 26, 2016: http://www.cpms.osd.mil/expeditionary/pdf/arcent%20rr%20sop%20final%20-%20jan%20%2011.pdf

U.S. Army Combined Arms Support Command, *Water Planning Guide: Potable Water Consumption Planning Factors by Environmental Region and Command Level*, Fort Lee, Va.: Force Development Directorate, November 25, 2008. As of April 24, 2015:
http://www.quartermaster.army.mil/pwd/publications/water/
Water_Planning_Guide_rev_103008_dtd_Nov_08_(5-09).pdf

U.S. Army Europe, *The USAREUR Blue Book: Base Camp Baseline Standards—A Guide to Base Operations Downrange*, undated. As of January 26, 2016:
http://www.aschq.army.mil/gc/files/Blue%20Book.doc

U.S. Army Europe and Seventh Army, *Base Camp Facilities Standards for Contingency Operations*, February 1, 2004. As of April 24, 2015:
http://www.eur.army.mil/pdf/Red_Book.pdf

U.S. Army Installation Management Command, *Installation Management Command Campaign Plan 2012–2020*, version 4.0, November 2011. As of January 26, 2016:
http://www.imcom.army.mil/Portals/0/hq/about/campaignplan/IMCP_18_Oct_11.pdf

U.S. Army Natick Soldier Research, Development and Engineering Center, Department of Defense Combat Feeding Directorate, *Operational Rations of the Department of Defense*, Natick Pamphlet 30-25, 9th ed., August 2012.

U.S. Army Natick Soldier Research, Development and Engineering Center, Expeditionary Maneuver Support Directorate, *A Guide for Tactical Training Bases: Shelters Handbook*, 2014.

U.S. Army Training and Doctrine Command, *The United States Army Concept Capability Plan for Army Base Camps in Full Spectrum Operation for the Future Modular Force 2015–2024*, Pamphlet 525-7-7, December 7, 2009. As of January 26, 2016:
http://www.tradoc.army.mil/tpubs/pams/tp525-7-7.pdf

———, *Integrated Capabilities Recommendation for Base Camp Engineering*, final draft, September 2012.

———, *Integrated Capabilities Recommendation for Base Camp Strategic Integration*, June 25, 2013.

———, capability manager for maneuver support, U.S. Army Training and Doctrine Command, "TECD 4a QIPR MSCoE Update," briefing to unknown audience, January 29, 2014.

U.S. Central Command, *Construction and Base Camp Development in the USCENTCOM Area of Responsibility—"The Sand Book,"* Regulation 415-1, April 15, 2009.

U.S. Code, Title 10, Armed Forces, Subtitle A, General Military Law, Part IV, Service, Supply, and Procurement, Chapter 169, Military Construction and Military Family Housing, Subchapter I, Military Construction, Section 2805, Unspecified Minor Construction. As of August 27, 2015:
http://uscode.house.gov/view.xhtml?req=granuleid:USC-prelim-title10-section2805&num=0&edition=prelim

U.S. Senate Committee on Appropriations, Subcommittee on Defense, *Statement by the Honorable John M. McHugh, Secretary of the Army, and General Raymond T. Odierno, Chief of Staff, United States Army, Before the Senate Committee on Appropriations Subcommittee on Defense, First Session, 113th Congress, on the Posture of the United States Army*, May 22, 2013. As of January 26, 2016:
http://www.appropriations.senate.gov/imo/media/doc/hearings/
2013%20Army%20Posture%20Hearing%20Statement.pdf

Vavrin, John L., William T. Brown, and William J. Stein, "USACE Support to Contingency Base Energy Management: Lessons Learned," Construction Engineering Research Laboratory, U.S. Army Corps of Engineers, May 10, 2013, not available to the general public.

Yeung, Douglas, and Margret T. Martin, *Spiritual Fitness and Resilience: A Review of Relevant Constructs, Measures, and Links to Well-Being*, Santa Monica, Calif.: RAND Corporation, RR-100-AF, 2013. As of April 18, 2018:
https://www.rand.org/pubs/research_reports/RR100.html